I0094962

Forests as Fuel

Forests as Fuel

Energy, Landscape, Climate, and Race in the U.S. South

Sarah Hitchner, John Schelhas,
and J. Peter Brosius

LEXINGTON BOOKS
Lanham • Boulder • New York • London

Published by Lexington Books
An imprint of The Rowman & Littlefield Publishing Group, Inc.
4501 Forbes Boulevard, Suite 200, Lanham, Maryland 20706
www.rowman.com

86-90 Paul Street, London EC2A 4NE

Copyright © 2022 by The Rowman & Littlefield Publishing Group, Inc.

All rights reserved. No part of this book may be reproduced in any form or by any electronic or mechanical means, including information storage and retrieval systems, without written permission from the publisher, except by a reviewer who may quote passages in a review.

British Library Cataloguing in Publication Information Available

Library of Congress Cataloging-in-Publication Data

Names: Hitchner, Sarah, author. | Schelhas, John, author. | Brosius, J. Peter, author.
Title: Forests as fuel : energy, landscape, climate, and race in the US South / Sarah Hitchner, John Schelhas, and J. Peter Brosius.
Description: Lanham : Lexington Books, [2022] | Includes bibliographical references and index.
Identifiers: LCCN 2021051330 (print) | LCCN 2021051331 (ebook) |
    ISBN 9781793632340 (cloth ; permanent paper) | ISBN 9781793632364 (paperback) |
    ISBN 9781793632357 (ebook)
Subjects: LCSH: Fuelwood—Southern States—Public opinion. | Climatic changes—Public opinion. | Forests and forestry—Southern States. | Fuelwood industry—Social aspects—Southern States. | African Americans—Southern States—Economic conditions. | Southern States—Social life and customs.
Classification: LCC TP324 .H58 2022 (print) | LCC TP324 (ebook) |
    DDC 333.95/3970975—dc23/eng/20211109
LC record available at https://lccn.loc.gov/2021051330
LC ebook record available at https://lccn.loc.gov/2021051331

For Naia Bulan, who has more energy than all the pine trees in Georgia, and for Rad, who gives us hope that the future is in good hands.
—Sarah Hitchner and Pete Brosius

For Susie, for her support and tolerance during fieldwork and writing.
—John Schelhas

Contents

List of Figures

Foreword

The lenses of time and experience have shaped the U.S. South and one of its sons. From a working-class Black community in a once-thriving mill town to the halls of a congressional testimony on extreme weather events, my story is a unique patch in the quilt of racial reckoning, climate change, and the natural system. Quilting is a southern tradition that weaves intricate patterns and designs to create stories of its people and provide warmth for them. As I consider *Forests as Fuel: The Intersection of Energy, Landscape, Climate, and Race in the U.S. South*, such images come to mind.

The South has been at the nexus of some of the most poignant transitions in history. Indigenous peoples were forced from their land along a Trail of Tears. Georgia and other colonies forged their independence through revolution. Kings, queens, and their people were stolen from their land and ferried like cargo to toil for others. That system birthed further turmoil on the road to justice—civil war, Jim Crow, the Civil Rights movement, and so forth. Within this backdrop of a complex human condition, something was also constant. The natural landscape of the South is as beautiful and vital now as it was during those periods of history. Yet now, the forests, rivers, and air are under a different stressor of our own making.

Climate change is a crisis for humanity and the natural system. Scientific literature has always been clear that the naturally varying climate system now has the imprint of people on it. While polar bears have been an enduring symbol of climate change, our way of life very much hangs in the balance. Agricultural productivity, public health, infrastructure, transportation, and economic vitality are directly impacted by the climate crisis. While everyone will feel the brunt of climate change, people of color and the poorest communities are disproportionately vulnerable.

Wood-based bioenergy is one of several solutions that offer hope for the mitigation of greenhouse gas emissions. Both mitigation and adaption are pathways forward, though some even propose risky geoengineering interventions. The solution matrix must, however, ensure that climate change is not "fixed" on the backs of marginalized or impoverished groups, particularly people of color.

The body of work presented here tells the story of wood-based bioenergy from the pillars of energy, landscape, climate, and race. With the American South as the backdrop, this necessary collision of anthropology, scale, energy policy, history, and culture is a vital contribution to the literature. It also triggers memories of the many Georgia days that I, a Black child, spent exploring the forests, asking science questions, and wondering how we impact it all.

Dr. J. Marshall Shepherd
Georgia Athletic Association Distinguished Professor
Director, Atmospheric Sciences Program
University of Georgia
Member, National Academy of Engineering, National Academy
of Sciences, American Academy of Arts and Sciences

Acknowledgments

Many people encouraged, supported, and facilitated the research that comprises this book. Our initial foray into what was often termed the "social acceptability" of bioenergy developed through conversations with Bob Rummer of the Southern Research Station of the USDA Forest Service. Enthusiasm for bioenergy development was very high in the scientific and forestry communities, but experience on the technical side of bioenergy development had made apparent that more nuanced understandings of the social issues involved were critical. The Southern Research Station provided initial funding for the literature review and field work that allowed us to gain a good sense of the range of responses of community members and landowners to bioenergy development. This research enabled us to compete successfully for a grant from the Sustainable Bioenergy Challenge Area of the USDA National Institute of Food and Agriculture—Agriculture and Food Research Initiative (NIFA-AFRI) (USDA-NIFA Award No. 2012-67009-19711), which supported in-depth research. In particular, we would like to thank Fen Hunt, the program manager for this grant, for her interest in an approach rooted in multisited ethnography rather than the more common questionnaire-based research.

While conducting our bioenergy research, we began to work with the U.S. Endowment for Forestry and Communities' Sustainable Forestry and African American Land Retention Program (SFLR). This outreach program focused on the issues of insecure land tenure and under-engagement in forestry that had long been a concern among Black landowners in the South, and we were able to add a research component that supported the program. This program had some overlap with bioenergy development, and it enabled us to engage further with racial issues associated with forest landowners. In particular, we would like to thank Alan McGregor, Sam Cook, Jennie Stephens, Alton

Perry, William Hodges, Alex Harvey, Cassandra Johnson Gaither, Rory Fraser, Viniece Jennings, Amadou Diop, and Puneet Dwivedi for their many contributions and efforts that facilitated our work. We developed further research on Black landowners and forestry in Georgia in collaboration with Puneet Dwivedi at the University of Georgia and Marc Thomas at Fort Valley State University with funding from the USDA NIFA Southern Sustainable Agriculture Research and Education Sub-award LS17-281, and this book also incorporates findings from this work. We also greatly appreciate the grant coordination support from the Owens Institute for Behavioral Research and academic support from the Center for Integrative Conservation Research at the University of Georgia.

This book is based on hundreds of conversations with landowners, community members, foresters, state and federal natural resource professionals, local government officials, industry employees, small business owners, educators, and representatives of nongovernmental organizations. They welcomed us into their professional and personal worlds in many different ways and generously shared their experiences, opinions, and feelings regarding bioenergy, forestry, and community. We are grateful for the time we were able to spend with them, and we hope we have represented them well. We also hope that our research has fulfilled its goals of elucidating the complexities of implementing wood-based bioenergy initiatives in local communities and offering suggestions for sustainable and equitable paths forward as renewable energy technologies continue to develop.

Chapter 1

Bioenergy Stories at the Intersection of Energy, Landscape, Climate, and Race in the U.S. South

ONE BIOENERGY STORY

On November 6, 2007, all eyes in the national and international bioenergy world were on the official groundbreaking for the new Range Fuels plant in Soperton, Georgia, which was expected to produce 40 million gallons per year of cellulosic ethanol using southern yellow pine as a feedstock. Incorporated in 1902, Soperton, Georgia, like many southern towns, is a forest-dependent community; it has long relied on forestry products (i.e., turpentine, pitch and naval stores, lumber, pulp and paper, pinestraw) as a main source of revenue for landowners and source of employment for community members. Soperton is located in the heart of the "pineywoods and wiregrass" or longleaf pine ecosystem that used to be the dominant ecosystem of the southeastern United States. The identity of the community is rooted in forestry; Soperton is known as the Million Pines City, a moniker that alludes to its claimed history as the birthplace of intensive silvicultural practices, including intentional planting of pines as an agricultural crop starting in 1926. In March 1933, *The Soperton News*, a weekly publication, became the first newspaper to be printed on paper made from pine pulp; a copy of this first edition is now housed in the Smithsonian Institute in Washington, DC. Soperton is very small (the town limits include just over three square miles), and its population hovers at just over 3,000. Built at the crossroads of railroad lines, like most southern towns, it was once much more prosperous and economically and socially vibrant than it is now. Since the heyday of pine, roughly the 1950s through 1990s (Carter et al. 2015), mills have closed, and the town and county have spiraled into economic decline. Now outmigration and poverty in the town and county, exacerbated

1

by long-standing racial inequalities, are evident; many people there with whom we spoke lamented: "Even our Dairy Queen shut down."

But Range Fuels could change everything. In Soperton, the initial announcement of the plant was met with great enthusiasm, as it would bring jobs for locals, an influx of money into the city and county from spin-off businesses, new markets and higher prices for landowners' wood products, and good publicity for Soperton. The 2007 groundbreaking ceremony was attended by a number of high-ranking government officials including then-U.S. secretary of energy Samuel Bodman and then-governor Sonny Perdue, who emphasized the state's commitment to renewable energy development by proclaiming Georgia as the nation's "Bioenergy Corridor." It is difficult to overstate the excitement among proponents of bioenergy both nationally and worldwide about Range Fuels scaling up its gasification technology (which had been successful at its pilot plant in Colorado) to commercial scale in Soperton. Commercial-scale production of cellulosic ethanol made from trees, mostly forest residues or woody materials left behind after logging operations, was going to be a breakthrough technology, revolutionize the energy landscape, create new markets for products that people wanted to get rid of anyway, and again position Soperton as a leader in forest-based innovations.

Fast forward a few years to another well-attended public ceremony at the Range Fuels plant, an open house in which the company opened its doors to show off the facility, its processes, and its products. Everyone there saw the big fuel trucks roll by in a parade-like display. One local man, a prominent landowner and businessman, said to an employee of Range Fuels, "I think that truck is empty. That trailer is bouncing way too much," to which the man replied, "Now, George,[1] you know there ain't no damn ethanol in there."

The "dog and pony show" that day was a grand gesture, and it excited promoters of "green energy"—in the bioenergy industry, forest industry, scientists, policy-makers—not just in Georgia but nationally and internationally. But the fissures in the dream were already evident to people on the ground. Despite receiving over $400 million in grants, loans, loan guarantees, and investor funds, due to technical problems in scaling up the technology, Range Fuels was only able to produce one batch of methanol from synthesis gas (which was possible in the 1920s and is a common process today). In 2011, Range Fuels declared bankruptcy and closed. The implications of Range Fuels' failure have been profound, both within the national biofuel industry and within the communities in and around Soperton. The story of Range Fuels, set in Georgia, called "the Saudi Arabia of biomass" by Susan Varlamoff, then-director of the Office of Environmental Sciences at the University of Georgia, in 2006, is one of many that tell part of a bigger story about the development of the biomass industry and the effects it has had on local communities throughout the region.

SOME BACKSTORY

This story, like others involving wood-based bioenergy all around the world, requires backstory. The early 2000s, in the years before and after the 2006 release of Al Gore's documentary *An Inconvenient Truth*, were a period of technoscientific optimism for the promise of green energy—especially bio-energy—as an antidote to the consumption of fossil fuels. Bioenergy was touted as a way to simultaneously meet multiple objectives including rural development, energy independence, and climate change mitigation. While more sobering assessments and critiques of the promise of bioenergy soon appeared, raising issues of food versus fuel among others, the exuberance, enthusiasm, and sense of hope that accompanied the promise of bioenergy reached its zenith.

The history of the fields of development, conservation, and sustainability are littered with seemingly endless cycles of disillusionment, as one idea or another is touted as a panacea (Ostrom et al. 2007), only to be reality-checked when the hubris of models encounters the complex realities of diverse local contexts. Bioenergy is no exception. Enthusiasm for bioenergy in the mid-2000s—manifested in national energy imaginaries and in local initiatives like Range Fuels—was driven by a combination of naïve tech-noscientific optimism and abstract economic models. It soon became clear, however, that the real impacts were messy and complicated. Understanding this messiness begins with the recognition that different actors (i.e., biofuel producers, fossil fuel distributors, environmental NGOs, wood products industries, community leaders) promulgate differential framings of wood-based bioenergy using certain discourses and narratives (i.e., national secu-rity, renewability, job creation) strategically. These framings are discussed and translated in different social contexts and on different spatial and temporal scales, and they intersect on the ground in specific locations with very real consequences for landscapes and communities. In this book, we will both share some of the stories in the southeastern United States that we have learned about from community members who have experienced them firsthand and also some of the broader political, historical, and social context in which these stories have played out. Offering ethnographic examples and analysis based on several years of multisited fieldwork in the region, mostly concentrated in two sites in Georgia and one in Mississippi, we use this book as an opportunity to explain some of the backstory behind the some-times dramatic headlines about bioenergy and to capture the challenges we encountered in studying a constantly shifting landscape of bioenergy development in the U.S. South.

BOOK STRUCTURE: PILLARS AND LENSES

Anybody who has ever taken an introductory anthropology course has no doubt heard that a core concept in ethnographic research is holism; that we anthropologists do not generally study cultural phenomena in isolation, but that we contextualize them with reference to other cultural elements. In a similar vein we found that in studying wood-based bioenergy, as John Muir famously wrote, "When we try to pick out anything by itself, we find it hitched to everything else in the universe." Likewise, we cannot tell the story of bioenergy development in this region in isolation. It is deeply connected to many other issues, and we have chosen what we consider to be the four main "pillars" which support this Parthenon—energy, landscape, climate, and race—and the next four chapters trace this. Chapter 2 introduces the different types of wood-based bioenergy products such as wood pellets, cellulosic biofuels (ethanol and biocrude), and combined heat and power (CHP) products currently and potentially produced in the region. We analyze the different trajectories of production and consumption these take, highlighting the fact that there is no one "bioenergy story" to tell here and that the language used to talk about bioenergy circulates among various actors strategically. In chapter 3, we ground the story of wood-based bioenergy in the U.S. South in the physical, social, political, and cultural landscapes of the region, exploring the layers of history literally embedded in—or eroded from—the soil and the ways that the history and potential futures of this heavily forested landscape are linked to energy creation and development. We include stories about what happens in local communities when bioenergy projects, especially those relying on experimental technology, fail, as well as stories about bioenergy facilities that have been successful, at least by industry standards. Chapter 4 focuses on the varying perceptions of and reactions to climate change that we encountered during our ethnographic research in the rural communities and at regional bioenergy conferences and workshops. Although climate change is a major driver of bioenergy policy, belief in and concern about it is far from universal among people in the U.S. South; concerns about climate often fall along partisan and racial lines. In chapter 5, we explore the complex racial dynamics of the region, noting the historical and current racial climate in which we conducted research. The U.S. South, and the country as a whole, has experienced a great loss of Black-owned[2] land since Reconstruction, and many Black landowners and community members feel uninformed about, or excluded from, current wood-based bioenergy developments in rural areas. Focus on these four elements as the pillars which support our exploration of wood-based bioenergy development, and the ways that they are intricately and uniquely intertwined in the U.S. South, add dimensionality to bioenergy's backstory. These pillars are integral to the multiple trajectories that this history of development has taken and continues to take in this region.

We connect these issues through three "lenses" which provide the internal structure for each chapter and together form the conceptual framework that holds the book together: imaginaries, scale, and translation. The concept of *imaginaries* draws our attention to visions of how things could be or ought to be. When we talk about the main pro-bioenergy imaginary, we are talking about a vision in which wood-based bioenergy promotes energy independence, enhances forest health, strengthens local wood markets, mitigates climate change, and benefits rural communities. As we will show, when this imaginary intersects with the reality on the ground, things can get messy. We show how various visions of bioenergy development have (or have not) led to real, observable changes in economies, landscapes, and intra-community dynamics (or illustrated extant complexities in profound and obvious ways). The concept of *scale* draws our attention to the spatial and temporal dimensions of bioenergy development (international policy, national energy discussions, and within local communities), and the dynamics and interplay between them, as various discourses and narratives about bioenergy emanate from different actors and circulate among and between different scales. The concept of *translation* draws our attention to the ways these various strains of discourse are used strategically to pursue different actualizations of imaginaries and the struggles that occur when actors try to make different perspectives mutually intelligible across difference. As we examine these dynamics, our conception of translation foregrounds both what has been referred to as the politics of translation (Spivak 1986) and incommensurability, recognizing that some values and goals are fundamentally incompatible and accepting that not all values and end goals can be accommodated in any one reality. Bioenergy development requires that real, hard choices must be made and acknowledgment that these choices benefit some people, economies, and landscapes at the expense of others. We examine what happens when people talk past each other because the visions are just too different and how to productively move forward with the understanding that some visions and values are incommensurable. In the next section, we briefly explain the pillars and lenses, and in each chapter, we offer evocative examples from our research in various communities throughout the South that illustrate the interconnections between them.

The Pillars: Energy, Landscape, Climate, and Race

Energy

As noted, significant interest in bioenergy, along with other renewable energy options, began to develop in the United States and the European Union in the early 2000s. This new interest, which followed earlier attention during the 1970s energy crisis, was related to rising gasoline and natural gas prices, concerns about overdependence on foreign oil, and growing awareness and concern about the role of fossil fuels in climate change (Aguilar and Garrett 2009; Ansolabehere and Konisky 2014; Strauss, Rupp, and Love 2013). Wood-based bioenergy was seen as an important and accessible part of a renewable energy portfolio, particularly in the southern United States, for several reasons. One, wood is an ancient source of energy and has continued to be the leading source of renewable energy in numerous countries in the Global North (Aguilar and Mabee 2014). Two, biomass-based liquid fuels represent one of the only options for transportation fuels, in the short to medium term, that can meet future metrics of environmental, social, and political sustainability (Brown and Brown 2012). Three, the southeastern United States has extensive timberlands, a large and established forest product industry and infrastructure, and excess capacity due to downturns in pulpwood markets, and is seen as ideally suited for wood-based bioenergy for power generation and liquid transportation fuels (Mayfield et al. 2007; Wear et al. 2010).

Policies in both the European Union and the United States have promoted bioenergy development. In the European Union, a series of energy directives issued from 2009 to 2017 mandated that 20% of each country's energy portfolio come from renewable sources (with the percentage increased to 32% by 2030 in 2018), with woody biomass playing a role in meeting this target (Lantiainen et al. 2014). A wood pellet industry developed in the United States in response to European Union renewable energy targets (Aguilar 2012) and subsidies for electricity production. In the United States, the 2007 Energy Independence and Security Act (EISA) set ethanol targets that included phasing in increasing quantities of biofuels made from cellulosic feedstocks (Dwivedi and Alavalapati 2009). To meet this target, cellulosic bioenergy development was aggressively promoted by the U.S. Department of Energy and other federal agencies (U.S. DOE 2006). Additional incentives in agriculture, rural development, and forest sectors also supported these goals (Lantianen et al. 2014), reflecting the fact that promotion of bioenergy was driven by efforts to simultaneously address climate change, promote rural development, and achieve energy independence and security (Bracmor 2015; Mayfield et al. 2007).

Developing a viable wood-based bioenergy industry in the southern United States, as with efforts to advance sustainable bio-economies elsewhere, clearly involves social as well as technological change (Mayfield et al. 2007;

McCormick and Kautto 2013; Delina and Janetos 2018). Social issues have become increasingly visible as bioenergy has been linked to rural development and is seen as a way to diversify the economic base of wood-dependent rural communities in the southern United States (Mayfield et al. 2007). Local economic development interests, supported by local media, have promoted bioenergy development as an alternative compatible with the existing forest product industry with important benefits for forest health (Mayfield et al. 2007; Bailey, Dyer, and Teeter 2011; Dyer, Singh, and Bailey 2013). While a variety of types and scales of wood-based bioenergy development exist, including home/community heating and co-generation of electricity, national and international energy policies led to the emergence of wood pellets for electrical power and advanced biofuels as leading near-term options in the southern United States in the early 2000s. McCormick and Kautto (2013) find that the ultimate goal of a new wood-based bioenergy system is neither the only renewable energy option nor achievable through a technological fix, but rather requires broad attention to sustainability and governance issues. To this we add that a deep and nuanced understanding of the cultural and ecological landscapes in which these bioenergy advancements are happening is crucial, as the history and ecology of a region shape every new development. As in any good southern novel, the physical, social, and cultural landscape itself is not a backdrop, but a central character.

Landscapes and Forests

The cultural landscape of the United States South is a palimpsest of many layers of human history, and the history of rural forested landscapes is rife with stories of abandoned farmland and a long history of forest products. Since colonization displaced Native American populations, land use in the South can be characterized as having three distinct eras: the pre-Civil War era of slavery and plantations, the post-Civil War agrarian society based on sharecropping and Black subjugation (e.g., Jim Crow), and the rise of modern industrialization and large-scale forestry (Aiken 1998; Carter et al. 2015). Outside the agricultural regions, native forests were heavily exploited in the early 1900s for both naval stores and timber, leaving a cutover landscape with little thought to future forests (Carter et al. 2015). By the 1930s, unsustainable farming and exploitive timber harvesting had left large areas of degraded land (Wear 2002). Over the course of the twentieth century, many of these cleared and cutover areas returned to pine forests either through natural regeneration or tree planting, and the South became the dominant timber-producing region in the United States (Wear 2002). There were a number of drivers of this change. Farming became more mechanized and concentrated on better lands, industrialization drew people out of rural areas, the pulp and paper industry developed, and industry and government programs promoted tree planting (Carter et al. 2015; Hart 1980; Kirby 1987; Rudel 2001; Wear 2002). Forest industries grew significantly after World War II, but some declines were seen at the turn of the twenty-first century as timber output has decreased periodically due to slowdowns in new home construction.

Even as the region has experienced a long-term decline in the pulp and paper industry (figure 1.1; see also Wear et al. 2013; Brandeis and Guo 2016), outputs of timber and other wood products remain high. A study released by the USDA Forest Service in 2014 noted that forests in the thirteen states comprising the U.S. South produce 63% of nation's total timber harvest; although they cover only 2% of the world's forested area, they produce 12% of the global roundwood supply and 19% of the world's pulp and paper supply, making it home to the world's most productive industrial forests (Oswalt et al. 2014). While many pulp and paper mills and other forest products industries have folded, leaving economic vacuums in many small towns, the industry as a whole has grown, becoming ever more efficient and more consolidated in the hands of large corporate forest owners such as Plum Creek, Georgia Pacific, John Hancock, and Weyerhauser.

Since the 1970s, woody biomass for energy production has been discussed and incentivized in the United States as a way to reduce dependence on foreign oil and broaden timber markets (Mittlefehldt 2016), and in recent decades, focus on it has intensified (Alavalapati et al. 2013). The 2008 U.S. Farm Bill created a new program called the Biomass Crop Assistance

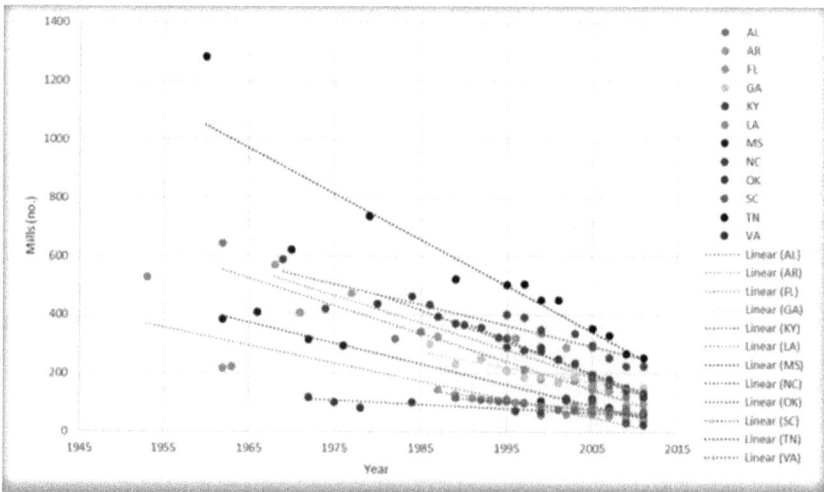

Figure 1.1   Decline in Pulp and Paper Mills in the U.S. South, 1945–2015, Showing the Sharpest Declines in NC, VA, and TN. *Source*: Credit: USDA Forest Service 2016.

Program (BCAP), which provided financial assistance to farmers and forest landowners for supplying biomass to commercial-scale bioenergy facilities, reducing the risk inherent in trying new crops, tree species, or harvesting rotations. The forested Southeast became seen as a vast resource to be exploited for this homegrown energy, and it has often been assumed that, regarding markets for woody biomass, the raw material for wood-based bioenergy, "if you build it, they will come." Some landowners told us that they will sell wood if markets are favorable, but our interviews with landowners concur with previous work on forest landowner decision-making that suggests that willingness to provide biomass is not simply an economic decision. The literature on family and nonindustrial private forest (NIPF) landowners, who represent 95% of the private forest owners and 63% of the private forest land in the South (Schelhas et al. 2003), makes clear that many southern forests are managed for multiple purposes. Butler and Leatherberry (2004) found that timber production was not among the top three reasons for owning family forests in the South, and that only 18% of family forest owners, who own 42% of the timberland, had harvested timber in the past five years. Newman and Wear (1993) found that NIPF owners place a higher value on standing timber than industrial owners, indicating that nonmarket benefits play an important role in their harvesting decisions. Forest lands are used for hunting and recreation, and they have important family legacy and aesthetic benefits to their owners (Bengston et al. 2011; Butler and Leatherberry 2004). This is particularly true for Black families, many of whom can trace ownership of the land to Reconstruction, and in some cases, even times of slavery.

While many wood-based bioenergy proponents claim that these budding industries will fill the gaps left by mill closures and that new markets will encourage landowners to plant more trees, many environmental organizations, health organizations, and environmental justice organizations in the region have warned against large-scale bioenergy development in the South. Some common concerns include the perception that local communities face threats from the wood energy industry, including forest destruction, noise pollution, degradation of air and water quality, decreased property values, negative impacts on local forest recreation and tourism, and increases in dangerous truck traffic in rural areas.

A major concern, particularly among environmental organizations, beyond the ecological threats caused by bioenergy production, transportation, and distribution, is about the feedstocks themselves. For example, the U.S.-based National Wildlife Federation (NWF) acknowledges that "bioenergy is one homegrown source of renewable energy that could help meet some of our energy needs," but cautions that "it must be produced in a way that has long-term economic viability, helps address climate change, and protects and enhances habitats and ecosystems" (Glaser and Glick 2012, 1). NWF claims that "Future bioenergy development should encourage ecological restoration and improve wildlife habitat through the use of ecologically beneficial biomass feedstocks such as waste materials and sustainably collected native plants and forest residues" (Glaser and Glick 2012, 2). They are concerned about the escape of exotic, invasive, and/or genetically modified organisms (GMOs) that are current or potential bioenergy feedstocks, especially energy grasses such as Miscanthus, Napier grass, and giant reed, as well as trees such as Eucalyptus. Although several commercialized versions of these species have been modified via a gene slice that is supposed to restrict their ability to reproduce, some question whether proof of their sterility has been adequately demonstrated (Glaser and Glick 2012, 18–20). NWF applies the phrase "the myth of sterility" in their widely distributed report on bioenergy as a way to influence the public to question the industries' claims that sterile hybrids will not pose any harm to native ecosystems. The Dogwood Alliance, an environmental NGO based in Asheville, North Carolina, has been particularly vocal in expressing dire warnings about the effects of wood biomass production on the region's forests. As part of their *Our Forests Aren't Fuel* campaign,[3] they have stated, for example, that the bioenergy industry's perceived need to reduce lignin in trees via genetic manipulation "could open the proverbial Pandora's Box of unintended genetic mixing of laboratory and wild trees" (Quaranda 2008, 4). This could not only change natural trees, but also facilitate the spread of "super pests and diseases" and threaten the rights of landowners to even own the trees on their property (Quaranda 2008, 7).

Numerous other studies also demonstrate public concern around the world about GMOs, whether as food sources or energy feedstocks (Finucane 2002; Nap et al. 2003; Miles et al. 2005; Delshad 2010). NWF strategically employs "the myth of sterility" and the Dogwood Alliance similarly uses the metaphor of "Pandora's Box" to tap into widespread concerns held by the public about the invasiveness potential of genetically modified species and their potentially disastrous effects on local ecosystems. We discuss other evocative metaphors that we encountered during our fieldwork in chapter 2.

There are many forest-dependent communities in the South, and, as with Soperton, often their community identity centers on this history of forestry. This landscape, however, is changing, due to widespread development, out-migration, farmland abandonment, the rise (and occasional stagnation) of plantation forestry, the introduction of new—sometimes genetically modified—species into forests and farms, and climate change. Its history, present, and potential futures are imagined differently by various people that live within these communities and people that make decisions about it—as in most places around the world, these are not always the same people, and this fact causes friction.

Climate Change

The effects of climate change are already being felt by communities around the world. Sea levels are rising, threatening many coastal cities, storms are becoming more intense and frequent, global average temperatures are rising, and regional climates are changing in various ways (IPCC 2014). Populations already vulnerable to storms or any variability in ecological conditions, such as in regions with water scarcity or communities living along coastlines, are especially vulnerable to the myriad effects of climate change. These ecological fluctuations are on a trajectory to intensify, leading to cities and towns flooding and disappearing beneath the ocean and ecosystems becoming further fragmented and irreparably harmed. There is near consensus among climate scientists that in order to prevent global ecological disaster, average global temperature rise must be limited to 1.5° C (some models say 2°) above preindustrial levels and that the global net emissions must be zero by 2050. Carbon capture and storage (CCS) is an essential tool in meeting climate targets. The Intergovernmental Panel on Climate Change (IPCC) and the International Energy Agency (IEA) have also shown that CCS is one of the lowest cost options, by as much as half, compared to other emission reduction technologies (IPCC 2014). Technologies that capture more carbon than they produce, known as negative emissions technologies, are a vital part of this portfolio of tools that can help the world achieve this goal. Significant increase of use of these technologies, as well as improved efficiency of them, are essential, especially in places where the most emissions are currently being produced. While many extant and developing technologies can pull carbon from soil, water, rock, and seawater for utilization and storage, the Global CCS Institute (2019) and others have noted that Bioenergy with CCS (BECCS) is one of the few currently available that can operate at a large enough scale to make a substantial contribution to climate change mitigation.

While national and international policies emphasize the importance of adapting to and mitigating climate change, these policies are influenced by a number of different actors with different priorities and goals. Bioenergy and forestry industry representatives, environmental NGOs, scientists, landowners, and community members all circulate ideas about the role that wood-based bioenergy can play in either mitigating climate change through increased forest cover and reduced emissions and pollutants compared to fossil fuels, or in exacerbating it when variables such as the production, transportation, and consumption of bioenergy products are fully tabulated.

The issue of carbon and its relation to climate change is a major concern among environmental NGOs, and several of these that oppose using wood for energy particularly question "the myth of carbon neutrality"; these include, but are not limited to, the Dogwood Alliance, Greenpeace, Carbon Trade

Watch, Energy Justice Network, and the Partnership for Policy Integrity. Many have noted that while they support efforts to reduce use of fossil fuels, they question the use of wood as an energy source. Several major international environmental NGOs in Europe (or with branches in the European Union)[4] have collaborated and issued joint statements on wood energy, outlining their particular concerns:

1) The long "payback period," during which more carbon is released into the atmosphere than is sequestered by replanted trees, which is critical as we get closer to the "tipping point" in climate change;
2) inaccurate greenhouse gas (GHG) accounting systems that mistakenly assume zero net emissions from biomass energy;
3) destruction of forests in areas both close to and far from biomass facilities; and
4) the limited nature of sustainably produced biomass (FERN et al. 2012).

A joint white paper produced in 2012 by Friends of the Earth, Greenpeace, and the Royal Society for the Protection of Birds (RSPB) echo these concerns, specifically in reference to the 2012 U.K. Bioenergy Strategy which called for developments in bioenergy that will "deliver genuine carbon reductions that help meet U.K. carbon emissions objectives to 2050 and beyond" (RSPB 2012, 2). In 2009, five prominent environmental NGOs in Europe[5] wrote a joint letter to the director general for the Environment of the European Commission urging the European Commission to require legally binding sustainability criteria for producers of biomass used for heating and electricity similar to the legally binding criteria for liquid biofuels and to refuse to compromise by promoting voluntary standards (FERN et al. 2012).

The Dogwood Alliance, an especially vocal opponent of wood-based bioenergy in the United States, states that "cellulosic ethanol, when looked at under close scrutiny, is at minimum a false solution and, in the worst-case scenario, a disaster for our forests that will exacerbate global climate change rather than combat it" (Quaranda 2008, 1). Their report claims that increased demand for wood as an energy feedstock will lead to more forest destruction, "immense" water use for irrigation and refineries, water pollution, soil erosion, soil compaction, reduced soil fertility, and increased air pollution. In 2013, the Dogwood Alliance joined with seventy-five other environmental NGOs, including many local and regional NGOs as well as large international NGOs such as Rainforest Action Network, Environmental Working Group, and BirdLife Europe, to call for an immediate moratorium on using whole trees for pellet production and in utility-scale biomass facilities, including co-firing with coal in existing power plants (Dogwood Alliance 2013).

Greenpeace Canada is strongly opposed to using wood from natural forests as an energy feedstock, though they also emphasized the importance of scale in relation to the sustainability of wood-based bioenergy. They released a report in October 2011 (Mainville 2011, 3) that explained that the assumption

> on which bioenergy is premised—i.e., that woody biomass is infinitely available and that burning it is clean and carbon neutral—does not stand up to scientific scrutiny and needs to be revisited . . . . Forest biomass cannot and should not replace fossil fuels on a large scale.

The report does note that small-scale forest bioenergy can be sustainable: "Using mill waste and residue, such as sawdust and non-commercial wood chips, to replace fossil fuels for local, small-scale heating systems is the most efficient use of woody biomass" (Mainville 2011, 2). However, they claim that using waste from sawmills and logging residues is a gateway to using whole trees from large-scale clear-cuts of natural forests; they also state that "there is no such thing as 'waste' in a forest" and that large-scale extraction of all woody materials from a site has devastating and long-term effects on soil fertility, wildlife habitat, and the ability of forests to mitigate damage from acid rain. While they say that biomass-based energy is in no way worse than using fossil fuels, they advocate policies incentivizing energy conservation and wind, solar, and geothermal energy options. These debates about the carbon neutrality (or not) of bioenergy between bioenergy industries and environmental NGOs do share the premise that climate change is real, is happening, and is a problem.

However, though there is overwhelming scientific evidence supporting the notion of anthropogenic climate change, it is still considered an unsettled (or biased) science among many Americans, particularly in the rural South. Bioenergy representatives have stated in public meetings in the U.S. South that bioenergy development will happen "whether you believe in climate change or not." This framing ignores concerns about the issue, rendering them irrelevant. In places in the U.S. South where wood-based bioenergy is developing, it is happening relatively independently of concerns about climate change. Even without the consideration of climate change, wood-based bioenergy may boost the local wood product markets; this would be a fine outcome for many rural forest landowners. At the same time, many minorities are well aware of their increased vulnerability to both climate change and efforts to mitigate it, such as the development of wood-based bioenergy plants near their homes. These views are two of many that collectively exemplify the inextricable links between bioenergy, forests, climate change, and race, which we address throughout the book, with a specific focus on climate in chapter 4.

Race

Racial dynamics are deeply embedded in the history and politics of the U.S. South, and even written into the landscape itself. The history of racial subjugation, violence, dispossession, and disparities in the region directly affects differential valuations of forests, development policies, and climate change, as well as divergent framings of social and economic issues and visions of a sustainable and just future. Race intersects with the history of land use change and forests in the U.S. South in a number of ways, including lower levels of managed forestry among minority landowners (which decreases their ability to participate in bioenergy markets), widespread loss of land ownership by Native Americans and Black Americans, higher rates of insecure land tenure among minority landowners, and environmental justice concerns regarding the site locations of bioenergy facilities.

There has long been concern with low levels of Black landowner engagement in forestry and within the forestry profession (Payne and Theo 1971; Hilliard-Clark and Chesney 1985). Racial and ethnic minorities and limited resource landowners have often been underrepresented and underserved in forestry-related developments in the South (Schelhas 2002; Schelhas et al. 2003). In particular, this is true both for Black forest landowners (Schelhas et al. 2003) and Black employees in forest products industries (Bailey et al. 1996; Bliss et al. 1998). This is sometimes associated with a desire to get away from impoverished rural lives in which only low-level manual jobs in forestry were available to Black Americans (Bailey et al. 1996; Bettis and Bettis 2017; Bliss et al; 1988, Bliss and Bailey 2005; Bliss and Flick 1994). But it also reflects discriminatory practices that made forestry education at universities in the South, more advanced jobs in forestry, and financial assistance through government conservation programs unavailable to Black landowners (Bailey et al. 1996; Payne and Theoe 1971; Schelhas 2002).

The history of rural Black landownership in the South is rooted in farming. Historically, almost all Black-owned farms in the United States have been in the South (Wood and Gilbert 2000). Black farmers emerged from slavery into sharecropping and, when possible, farm ownership (Daniel 2013). However, Black-owned farmland peaked around 1920 at about 926,000 acres and dropped precipitously over the course of the twentieth century at rates higher than the decline in White-owned farms (Wood and Gilbert 2000; Zabawa 1991). While small farms declined most rapidly, and Black-owned farms tended to be smaller, Black-owned farms decreased at a higher rate even when controlling for farm size (Wood and Gilbert 2000). Over this same time period (largely beginning with World War I), in what is known as the Great Migration, millions of Black Americans moved from the South to northern (as well as western) cities for industrial and service sector jobs

(Trotter 1991). Kinship relations between migrants and their southern homes both facilitated the Great Migration and were maintained as migrants made regular visits south, and some families retained their southern farms against all odds (Gottlieb 1991, 72–73). This pattern continued until the 1970s, when return migration of Black Americans from the North to the South, often to urban areas but also to rural homelands, became dominant (Stack 1996). Some of these return migrants returned home to care for elders and/or take over the management of their family lands (Stack 1996).

For both longtime residents and return migrants, family land was often in heirs' property. Land was often passed on across generations without wills due to distrust of and disenfranchisement from the legal system (Zabawa 1991). Heirs' property is held in common by individual shareholders who each own a fractional interest in the entire property, which generally remains in a deceased owner's name (Dyer and Bailey 2008). Shared ownership in the form of heirs' property often makes it difficult to productively use land (due to requirements for clear title for timber harvesting and participation in government programs and to intra-family disagreements on land use) and reduces the wealth of affected families (Dyer and Bailey 2008). As the viability of small family farms declined, heirs' property and other land often became covered in unmanaged second-growth forest (Schelhas et al. 2017a). Yet, lack of engagement in forest management by Black landowners has persisted and is intertwined with the problem of heirs' property (Hilliard-Clark and Chesney 1985; Gordon et al. 2013). This leads to limited opportunity for minority landowners to benefit from wood-based bioenergy development.

In addition to likely exclusion from wood bioenergy markets for minority landowners, other racial concerns are an important and often overlooked aspect of wood-based bioenergy development specifically and forest-based rural development in general. For example, minorities and politically progressive community members have expressed concern about environmental justice implications of the siting of some bioenergy plants and unequal levels of social vulnerability to their impacts. Bullard (2005, 2011) raises particular concern about racial equity related to the development of clean, green, and renewable energy, including biomass in the South, noting that plants have often been sited in poor Black neighborhoods. This can expose these neighborhoods to toxic emissions from plants and to vehicle emissions from trucks bringing feedstock to bioenergy production facilities.

There are both real and perceived impacts of bioenergy development in rural and urban communities, forcing the question of who bears costs and who receives benefits of bioenergy industry development. While the decentralization of power (both literal and figurative) that often follows wood-based bioenergy development (because facilities are sited in rural areas with high concentrations of feedstock) is often touted as a democratic benefit to society, Mittlefelhdt (2016) explains that the reality is more complex. She (2016, 19) describes

> a core paradox of local power: distributing power away from the centralized systems of production that brought unprecedented amounts of energy to American consumers in the twentieth century required local communities to accept the ecological and public health risks of their consumption.

The decentralization of energy systems has illuminated not only the challenge of equitably distributing the costs and benefits of local energy production but also wades into complex social, economic, political dynamics of local communities; these can include disagreements over land use, unstable land tenure among vulnerable populations, and competing narratives by various actors such as forest products industries, rural development advocates, environmental activists, and local community members.

In the development of a bioenergy plant, like the implementation of any conservation or development project, certain groups often have more power to shape activities and, even when the overall results are positive, there are often winners and losers. An equitable approach to bioenergy development recognizes this inevitability but seeks to incorporate the perspectives of all stakeholders, strives to give voice to underrepresented perspectives, and looks for solutions that address the needs of underrepresented and minority stakeholders (Schelhas and Lassoie 2001). This is not only ethically desirable but also pragmatic because they are often vocal and respond in variety of different ways, including actions such as lawsuits or protests.

The Lenses: Imaginaries, Scale, and Translation

Imaginaries

Efforts to develop a renewable energy system based on wood-based bioenergy in the U.S. South can be analyzed through the concept of "sociotechnical imaginaries" (Jasanoff and Kim 2009, 2013), or visions of the way the world ought to be and the way it ought to work. Imaginaries can serve as powerful cultural resources that support and shape societal efforts to transition to new energy futures. Sociotechnical imaginaries are "collectively imagined forms of social life and social order reflected in the design and fulfillment of nation-specific scientific and/or technological projects" (Jasanoff and Kim 2013, 190). Eaton et al. (2014, 227–228) draw our attention to the concept of sociotechnical imaginaries for renewable energy technologies, including bioenergy derived from woody biomass, with their observation that imaginaries for bioenergy derive from state actors who envision a future where energy and economic interests will be met with homegrown resources . . . providing "green" means to address salient social problems such as the nation's dependence on foreign and domestic fossil fuel supplies, climate change, pollution, environmental degradation, national energy security, and (rural) economic depression. The term *imaginary* connotes the way these visions provide an attainable end goal, or collective vision of a feasible, desirable future social order, provided by technological projects.

This view of bioenergy as addressing a set of multiple and loosely related issues that include energy independence and security, rural development, climate change mitigation and other environmental issues has been common in discourse across various types of wood- and crop-based bioenergy development in the United States (Burnham et al. 2017; Bain and Selfa 2013; Kulscar, Selfa, and Bain 2016). As such, it can be considered the dominant national bioenergy imaginary—connecting social and technological realms, promulgated by powerful interests, and embodying a clear moral and environmental vision of a future involving bioenergy.

Yet this bioenergy imaginary has not gone uncontested; different stakeholders promote or subscribe to different imaginaries, and they have different motives for doing so. One is the tendency to see biofuels as a scam, selling an unviable product to enrich its proponents (Hitchner, Schelhas, and Brosius 2016). Government subsidies for biofuels, ranging from those for the Range Fuels plant (Chapman 2012) to military spending on the Great Green Fleet, an effort by the U.S. Navy to develop alternatives to conventional fuels (Cardwell 2012), have been criticized as wasteful government spending. A second imaginary focuses on public health and environmental justice. Supplying pellets to Europe's wood-burning power-generating plants, often called "biomass incinerators" by opponents, is sometimes referred to as turning the United States into a European resource colony (Schlossberg 2013). Interpreting biomass power plants as incinerators calls attention to air pollution concerns related to burning wood, and it has raised environmental justice concerns when these plants are located near minority communities (Bullard 2011; Hitchner, Schelhas, and Brosius 2014). A third alternative imaginary revolves around ecological impacts. Questions about renewability and carbon neutrality have been raised (McBride 2011; Phillips 2015). Environmental groups have maintained that bioenergy threatens to push forests—valuable for sustainable forest products, tourism, and as cultural resources—to the brink of disaster by causing irreparable harm through deforestation and degradation (Quaranda 2008). Environmental and conservation organizations have expressed concern that bioenergy can have potential impacts such as soil erosion, decreased water quality and quantity, and conversion and deterioration of wildlife habitat in exchange for only modest greenhouse gas reductions (McGuire 2012).

These multiple imaginaries regarding bioenergy are all based in various scientific disciplines. But as we know, science has a long history of being used strategically; books have been written specifically about "how to lie" with statistics (Huff 1964), charts (Jones 1995), and maps (Monmonier 1991). As such, close examination of the motives of various actors, as well as careful reflection on their capacity to persuade others, is critical in gaining a deeper understanding about how ideas about bioenergy are formed and shared. As we will show, ideas about bioenergy are linked to broader cultural elements that both connect and transcend energy systems. As noted by Mittlefehldt (2016, 19),

> By examining the nuances of narratives constructed by those who promoted and those who opposed renewable energies like biomass, we gain a more complex view of the dynamic interaction between cultural ideas about nature, power, and the implementation of new technology.

In chapter 2, we more fully explain the types of imaginaries circulating about wood-based bioenergy and how various actors use specific language and linguistic mechanisms to persuade others of the benefits, and risks, of bioenergy development.

Scale

The issue of scale arises across epistemological divides; it is both an issue of salient concern in the natural sciences and in those academic domains that approach their subject from a critical perspective. Levin (1992, 1944), for instance, described scale as "the fundamental conceptual problem in ecology, if not in all of science," while the "politics of scale," a phrase coined by Smith (1984), explicitly acknowledges that scale is a social construction (Smith 1984, 1992; Delaney and Leitner 1997; Martson 2000; Brenner 2001; Herod and Wright 2002) and neither a predefined geographical territory nor politically neutral (Swyngedouw 1997). The issue of scale has always been an important component of ethnographic research (Marcus 1995; Li 1996; Brosius 2004; Witter et al. 2015), as it is commonly accepted among ethnographers that society (or any subset of society) is composed of multiple actors acting with multiple sources of knowledge and operating at multiple scales of governance (Marcus 1995; Olsson, Folke, and Berkes 2004). Multi-scalar research has emerged in recent decades as a way to add both spatial and temporal dimensionality to topics as complex as bioenergy (Scott et al. 2014).

In this book, we discuss the localized implications of global-scale international policies regarding energy, bioenergy, and specifically wood-based bioenergy on forest-dependent communities in the U.S. South. We draw from previous ethnographic fieldwork at global events such as the 4th World Conservation Congress (WCC) in 2008 and the 10th Conference of the Parties of the Convention on Biological Diversity (CBD CoP10 in 2010) and from research on other global discussions on bioenergy. We also examine how policies and events in other countries and regions have affected bioenergy development in the U.S. South; for example, the European Union renewable energy standards necessitate high quantities of imports of wood pellets sourced in the U.S. South. We examine how national policies on energy in the United States, such as subsidies for renewable fuel and power generation, fuel blend requirements, and tax breaks as incentives for renewable energy facilities affect local communities. We also explore the regional history, culture, and ecology of the southeastern United States, showing why this region is ground zero for wood-based bioenergy development and the implications that these developments have had on the ground. Finally, we examine the local scale of the communities which served as our primary and secondary ethnographic field sites across the U.S. South and examine the specific configurations of their particular political, social, cultural, and ecological histories in relation to bioenergy development. There are sometimes disconnects between the scales on which bioenergy developments are imagined, promoted, and enacted, and these disconnects reveal the conflicting dynamics that can occur between actors making decisions at different scales (Kuchler and Linnér 2012; Scott et al. 2014).

While attention to the scalar mismatch of costs, threats, and benefits is fundamental to economic models of large-scale bioenergy development, it also elucidates the imaginaries in which bioenergy is literally produced and socially enacted, providing the context necessary to understand the effects of locating commercial-scale wood-based bioenergy facilities in small, rural, economically depressed communities. In chapter 3, we show how various stakeholders spoke about different factors affecting potential threats and benefits in the context of the Range Fuels and LanzaTech bioenergy facilities in Soperton, Georgia. In doing so, we show that while global discourses promote bioenergy development as a mechanism for mitigating global climate change and as a vehicle for economic resilience in small and rural communities traditionally dependent on forest resources, local discourses are simultaneously circulating that often question climate science and resist government intervention in energy markets, for example, in the form of subsidies for new bioenergy companies.

We examine temporal scale in this book as well, particularly with regard to the four pillars we connect with bioenergy development (energy, landscapes, climate change, and race), and we place our fieldwork within the time frame in which we conducted it. We examine why wood-based bioenergy in the southeastern United States has emerged as a new use for the "wood-basket"—an extension of the breadbasket metaphor we heard often in our research—which has resulted, some argue, in a surplus of wood after large-scale conversion of farmland to forestland (especially following the path of the boll weevil that destroyed large-scale cotton production) and the closings of many local wood mills as a result of competition from large-scale facilities. We also locate our research within the frame of a time of convergence of policies and concern (or not) for climate change, support (or not) for transitions to renewable energy (however defined by various actors), and broader cultural attention to the current inequalities resulting from entrenched systemic racism on which our nation was formed. Finally, we explore the impacts of the historical rise and fall of energy markets, and the timeliness of new energy developments; these include widespread fracking initiatives, controversies over developments such as the Keystone XL pipeline, the ability of wood-based bioenergy products to finally compete economically with fossil fuels (with and without subsidies), and the prevalence of industrial and private, as well as federal, funds for research on technological innovations including biocrude oil and controversial products involving genetic modification and synthetic biology.

Translation

The topic of translation is one that for centuries has attracted scholarly attention and generated debate, and in the last century this legacy has continued across a range of fields: philosophy, linguistics, semiotics, hermeneutics, comparative literature, anthropology, and others. Recent debates have been informed by feminism, poststructuralism, postcolonial studies, indigenous studies, and the ontological turn, and we have witnessed the emergence of the new field of translation studies.

As anthropologists, we are particularly attuned to questions of translation. Since at least the time of Franz Boas, anthropologists, whose work is founded on a legacy of ethnographic understanding, have struggled with issues of translation (Hanks 2014, 17). It is not an exaggeration to say that ethnography is itself a form of translation writ large (Leavitt 2014, 194). As anthropologist Susan Gal argues, "By any definition of translation, anthropology and ethnography have always relied on it as part of method" (Gal 2015, 228).

While the overwhelming focus of scholarly work on translation across a range of disciplines has focused on issues related to questions about translation across languages (interlingual translation), there has also been a persistent strain of work that is concerned with questions concerning how translation functions *within* the context of a language community (intralingual translation). Tracing the genealogy of this distinction to the linguist Roman Jakobson, Gal (2015, 229) defines intralingual translation as "the rewording of verbal signs by means of other signs of the same language."

According to Hanks (2014, 18), "translation is a constant and unavoidable part of any single culture, and not only a problem of comparison." Commenting on what he terms *intracultural* translation, Hanks argues that it "plays a constitutive role in the social life of any human group, and not only in mediating between different groups and languages" (Ibid., 18). He argues that "speakers of any language routinely translate themselves and others in the same language" (Ibid., 21) and that "not only is translation a ubiquitous feature of ordinary monolingual speech, but the intralingual translation of an expression quite simply *is its meaning*" (Ibid., 21).

In this book, we draw on this fundamental distinction between two forms of translation to argue for a broader conception of translation that entails *any effort to speak across difference*. Such a perspective foregrounds the ubiquity of translation as a constant: between languages, between disciplines, between academics and fields of practice, between different sectors of any given society, and across scales. Viewed in this way, the task of translation is invoked in any context marked by cultural, political, epistemological, or ontological diversity irrespective of whether or not linguistic diversity is a factor. We follow Gal in recognizing that "translation in its broadest sense is the expression in one semiotic system of what has been said, written, or done in another" (Gal 2015, 227).

Following Spivak (1993), Asad (1986), Álvarez and Vidal (1996), and others (Fernández and Evans 2018), another element in our conception of translation is that the process is almost inevitably laden with power and that attention to the politics of translation is a necessary component of understanding contexts marked by power asymmetries, however such asymmetries may be constituted (i.e., race, class, gender, ethnicity, or others). Writing about the politics of translation specifically in the realm of policy-making, Clarke (2015, 44) stresses the importance of

> being attentive to the ways in which policies—as they move—may imagine and inscribe positions, hierarchies, relations of domination and subordination, and forms of power and authority. Policies do so even—or perhaps especially—when they come to speak in neutral, technical or administrative vocabularies. . . . If we are to take translation as a sensitising tool to direct attention to the multiple, the plural, the contradictory and the awkward in the policy process, then we need to see "unintended consequences," "unforeseen scenarios" and "unanticipated reactions" not as unfortunate by-products, but rather as systematic features and

likely consequences of policymaking in its conventional form. This, we think, is a necessary analytic commitment to a double orientation to the movements of policy and power: to recognise plans and projects, but to be attentive to their interruptions, disjunctions and failures.

In the context of this book, attending to the politics of translation allows us to engage with both the strategic and unintentional ways actors enmeshed in complex community assemblages give voice to, or contest, narratives and imaginaries as they navigate through cycles of enthusiasm and disillusionment of bioenergy.

Another element of translation we wish to foreground in this book is recognition that processes of translation are not merely dialogic, but that words, concepts, discourses and even stories travel and circulate widely through local, national, and international circuits. The impetus for this derives from scholarship on "friction," "words in motion," and "buzzwords" in the mid- to late 2000s (Tsing 2005; Gluck and Tsing 2009; Cornwall and Brock 2005; Buscher and Mutimukuru 2007). Tsing, for instance, traces how the story of Brazilian rubber-tapper Chico Mendes circulated through international rainforest activist circuits to Malaysia and Indonesia. Cornwall and Brock (2005) trace a series of global governance concepts in the field of development, including "participation," "empowerment," and "poverty reduction." In discussing what she terms "translation chains," Gal (2015, 231) notes that "translations . . . are never simply repetitions. They add the moral weight of others' voices to one's own, or alternatively, attribute responsibility for the utterance to others." Further, "As colonial transactions show, citation/translation can convert the typified speech of one group into evidence for another group's project. It can also do further metapragmatic work: justification, accusation, alliance, mobilization" (Ibid., 231). We address these issues specifically in chapter 2, where we provide examples of words, narratives, and metaphors about bioenergy that flow among various actors, either energizing or negating certain aspects of the pro-bioenergy imaginary.

Finally, one of the key elements of the politics of translation is the dynamics of commensuration and incommensurability. Theories of translation have a long legacy of commentary on translation as a process of commensuration and the challenges of conveying meaning across language spaces. The question is: at what point do we encounter the limits of commensurability in the process of translating words, meanings or metrics across zones of difference—linguistic, political, disciplinary, epistemological or otherwise?

Against the will to commensurate is recognition of the possibility of incommensurability (Kuhn 2012 [1962], Feyerabend 1962; Povinelli 2001). Our conception of translation extends toward recognition that whether we are concerned with words, meanings or metrics, translational struggles are ubiquitous and unavoidable, and we should not always assume that there is a path to commensurability. Incommensurability is particularly consequential when it is not acknowledged or recognized, or when the assumption is made that it can be overcome by an act of will.

WHY ETHNOGRAPHY?

Energy markets are notoriously volatile, whether due to supply shocks driven by weather events or international crises, spikes in demand, the adoption or loss of subsidies, or the disruptive effects of new technologies like fracking that destabilize the elegant supply and demand models theorized by economists and enacted in actual commodity transactions. Sometimes these volatilities and disruptions are obvious and visible; think long lines at gas stations during the 1973 oil embargo or the more recent "Bakken Boom" in North Dakota driven by fracking. But just as often they are diffuse, slow-motion and invisible, as cycles of prosperity and deprivation play out in the lives of those affected, and as energy imaginaries are reinforced or torn asunder through cycles of enthusiasm and disillusionment. Such cycles—of both markets and imaginaries—draw our attention to the disjuncture between the abstract models and financial calculations that drive so much planning by government and industry—as in where to build a new energy facility—and the lived experience of individuals and communities affected. This disjunction is both scalar and a dynamic embodiment of translation and incommensurability. We believe that seeing it and making sense of it is made possible by ethnography.

In the case of bioenergy, the abstract economic imaginaries of growth, prosperity, and profit rest on abstract assumptions about the feedstock supply chain and the willingness of individuals to supply biomass from forests they own or manage. What their abstractions overlook is the messy diversity of conceptions of what a forest is, and why forests are managed or maintained: one person's feedstock is another person's hunting land and yet another person's savings account for their children's education. This messiness is not a pathology. Rather, it expresses a politics of scale across which people encounter multiple incommensurabilities in their efforts to speak across difference. Ethnography exists to elucidate that messiness.

Ethnography is no longer just about what happens in local communities. Increasingly, it is multisited and multi-scalar, tracing links across global, national, and local contexts and processes. As anthropologists, we are committed to this vision of ethnography as a necessary corrective to a legacy of sometimes parochial near-sightedness. It certainly informs our approach to the study of bioenergy is the U.S. South. Industrial plants for bioenergy production, envisioned and planned in distant locales, have been proposed, sometimes constructed, and—with uneven success—operated at a commercial scale in specific communities. Ethnographic research is important to improve our understanding of how people in these communities envision and experience the cycles of development and disillusionment that have characterized bioenergy development initiatives. Ethnographic analysis of the cases presented here can elucidate various, often competing, worldviews and ways of discussing these interconnected issues. It can also help us address the question of how to promote renewable energy policy and production in ways that are beneficial to local communities most directly affected by these developments, especially in cases where bioenergy projects move forward even as the technology is not yet fully developed. Finally, an ethnographic approach to studying these stories—which lie at the intersection of sociotechnical imaginaries, circulating narratives, and concrete developments on the ground—helps us to identify areas of convergence and points of compromise in a tumultuous political climate characterized by division, confrontation, and polarization.

Funding for bioenergy research has traditionally been reserved for technological or biological research aimed at improving bioenergy technologies and feedstocks. Social science research has until recently mostly been limited to economic studies using rational choice behavior as a yardstick and "social acceptability" of bioenergy as an end goal. In 2010, when enthusiasm for wood-based bioenergy development was very high (Mayfield et al. 2007), we began a research project as a collaboration between the authors from the USDA-Forest Service Southern Research Station (SRS) and the University of Georgia Center for Integrative Conservation Research (CICR). This project focused on understanding the social context of biofuels and emerging bioenergy projects and options in the southeastern United States, as a complement to investments in technical research. We initiated an ethnographic study of communities and landowners around major bioenergy plants in the southeastern United States. We began with general research on bioenergy in the southern United States and intensive research around the Range Fuels plant in Soperton, Georgia, which at that time was a promising cellulosic ethanol plant using woody biomass as its feedstock. Based on our preliminary research in 2011/2012, we successfully competed for a U.S. Department of Agriculture AFRI-NIFA[6] grant in a program focused on socioeconomic analysis of biofuel development on rural communities (Sustainable Bioenergy Initiative). By the time we submitted our proposal, the Range Fuels plant had ceased to operate and was preparing for foreclosure sale.

We wrote our proposal to the USDA with the end goal of gathering a variety of perspectives and firsthand accounts from communities within which bioenergy development was actually happening. Though this type of research had never been funded by this program before, the National Program Leader for the Institute of Bioenergy, Climate, and Environment at NIFA believed that the ethnographic approach that we proposed could reveal some of the social complexity on the ground. Though we used safe language ("social acceptability") to write the proposal, once funded, we were allowed free range to conduct our research as we saw fit and to publish our results widely in disciplinary journals. We felt that ethnographic research was necessary to understand the experiences of community members and landowners in communities where bioenergy facilities are located, and we intend this book as proof that NIFA's faith in our approach was justified.

We offer this book with the caveat that this is only part of the story and that we did not conduct full-scale ethnography, in which case we would have become immersed in one community for a long period of time (defined by most anthropologists as one year or more). Rather, it is a multisited ethnography utilizing multi-scale analysis from multiple sources of information; we aimed for a regional scale analysis of wood-based bioenergy development rather than a deep analysis of one community. We have missed a lot, no doubt, but as all writers and researchers do across disciplines, we piece together what we do know in a way that we believe offers a compelling and nuanced story about what wood-based bioenergy development means on the ground in this particular region.

METHODS

Ideas on our methodology developed from long-term ethnographic research in other countries and within this region through trial and error, arguably the best way to learn what methods work for us as researchers and for this particular research project. Our conceptual framework combines theoretical trajectories developed in collaboration with other research teams over the past several decades (Schelhas and Pfeffer 2008; Schelhas and Lassoie 2001; Hirsch et al. 2010; McShane et al. 2011; Witter et al. 2015; Brosius and Campbell 2010; Campbell et al. 2014). The three coauthors are all cultural anthropologists with combined expertise in political ecology, cultural landscapes, land tenure, energy, and participatory research, and we have published widely on research related to the political ecology of conservation, minority landowners, social vulnerability and perceptions of climate change, and forested landscapes of the southeastern United States. This book draws mostly on ethnographic data collected during several years of fieldwork beginning in 2011,[7] the year that Range Fuels folded and that other technological innovations in other states in the region were still garnering much excitement and funding.

Lassiter (2005, 93) claims that ethnography is now often conducted in an "ever-changing, shifting, and multi-sited field." This is especially true of research on biofuels, which necessitates a multisited approach to studying the socioeconomic impacts of bioenergy development on a regional or even community scale. Ethnographic analysis of the cases presented here can elucidate various, often competing, worldviews and ways of discussing these interconnected issues. Our study of the process of envisioning and implementing sustainable bioenergy involved actors and discourses found in multiple sites, including both places and events, and therefore was well suited to multisited ethnography where people, connections, associations, and relationships are followed across space and time (Marcus 1995). In this interconnected world, field site boundaries are inherently arbitrary and defined by the researcher, and we chose to focus our research on the process of bioenergy development with ethnographic field research on the ground around new bioenergy facilities and at bioenergy events.

Using a complementary array of qualitative social science methods, we conducted ethnographic research in three communities in Georgia and Mississippi with different types of bioenergy facilities. We spent three months living in each of these three main field sites and interviewing many different stakeholders: landowners, community members, local development board members, school board members, local politicians, cooperative extension agents, loggers and others employed in the forest industry, and employees of bioenergy facilities. For these interviews, we employed a mix of both structured and semi-structured interview methods, during which we took detailed notes on both questions and responses and immediately transcribed them. We also transcribed fieldnotes about the location of the interview, relevant observations about the interviewee, and our reflections on the interview. We conducted a total of over 200 individual interviews for bioenergy-specific research, in addition to approximately 157 other interviews across the U.S. South for other research projects. These projects were primarily focused on identifying a range of Black perspectives on forest management, land tenure challenges, and social vulnerability to climate change; during these interviews, we discussed perceptions of forests, forest management objectives, beliefs about climate change, and knowledge about and participation in bioenergy markets. About 130 of these 157 additional interviews (and roughly 60 of the 200+ bioenergy-specific interviews) were with Black landowners. With another colleague (Cassandra Johnson Gaither of the Southern Research Station of the USDA Forest Service), we also conducted six focus groups with Black community members in downtown Atlanta in collaboration with organizations such as the Concerned Black Clergy, National Action Network, local NPUs (neighborhood planning units), and several AME (African Methodist Episcopal) and other predominantly Black churches in Atlanta. We also conducted nine focus groups in Spanish with Latinx community members in suburban Atlanta about issues related to social vulnerability and perceptions of climate change.

In our primary bioenergy research field sites, we participated in community activities and temporarily joined local organizations such as churches, a book club, a garden club, a minority community organization, a running group, and a public service club. At these events, we participated in ongoing group activities and introduced ourselves as researchers interested in interviewing community members; in this way, we met directly and were introduced to a number of interviewees. We kept field journals in which we recorded informal conversations with numerous other community members, as well as our personal reflections on people and events in the field sites. We also gave formal public presentations about our research project in each field site. We visited several additional sites with bioenergy facilities in Georgia, Louisiana, and Alabama and conducted interviews in these areas with extension agents, forest professionals, forest landowners, and employees of bioenergy facilities (see figure 1.2).

We also conducted event ethnography (Brosius and Campbell 2010) through attendance at a series of eighteen regional conferences and workshops on bioenergy[8] and participation in at least twenty-seven regional and national bioenergy-related webinars and conference calls. These events were mostly attended by bioenergy representatives and researchers working on technological advancement of bioenergy development. Ethnographic study of these events complemented our field site research, as we focused not only on

Figure 1.2   Coauthors (from left) John Schelhas, Sarah Hitchner, and J. Peter Brosius, 2013. *Source*: Credit: Sarah Hitchner.

the content presented during the sessions but also on the observable interactions between various actors. For example, we noted polite tension at times between representatives of environmental NGOs and bioenergy industry representatives, particularly when the former asked for clarification about potential negative impacts of burning biomass on human health and about ecological risks such as deforestation and damage to fragile ecosystems. We also noted how these meetings, which range from fully public to invitation-only, are utilized as venues for public announcements about new technological breakthroughs, biofuel facility openings, or developments in bioenergy policies. With very few exceptions, speakers at these events were either representatives of bioenergy industries or researchers sympathetic to industry; although there was some diversity of opinion represented at these events, we suggest that overall, they share and promote the idea that bioenergy is a promising pathway to energy independence through improved technology.

We view these events as an extension of community-based fieldwork in the field sites; the network of actors that attend these regional workshops and conferences could also be considered a "community." These were places where pro-bioenergy imaginaries were perpetuated and honed; meanwhile, the reality played out on the ground in the communities where bioenergy was actually produced. We wanted to capture the intersections and areas of divergence between the two. For example, our research in Soperton, Georgia, explored in detail in chapter 3, shows how the sociotechnical imaginary that promotes bioenergy as a driver of community resilience does not capture the risks and consequences of the potential failure of a bioenergy facility in a small, forest-dependent town, nor how it directly competes with a political imaginary in which government and business entities profit at the expense of communities and taxpayers regardless of whether a bioenergy project succeeds or fails.

We conducted research knowing community members would be reading our writings on it; that was always an explicit understanding, and we followed through on submitting our writings to any community member who accepted our offer of it. This knowledge has contributed to multiple conversations among the coauthors about how to portray the communities and the individuals within it accurately and fairly. Consistent with research protocol, we have provided pseudonyms of composite characters with lack of any identifying features, which is extraordinarily difficult in a small town where there may be only one or two "large-scale forest landholding families" or one or two main "Black community leaders." While we have had to be vague in describing the people with whom we spoke, we have retained names of cities, towns, counties, and companies, except in circumstances where that information might reveal the speaker. When we have provided direct quotations, those are not altered in any way, unless necessary to preserve confidentiality.

FIELD SITES

Energy markets are complex, multi-scalar, and always shifting as new technologies, new energy demands, and new constraints constantly disrupt whatever preceded them. The question that animates this research is what happens when these large-scale processes that occur beyond the purview of local contexts produce the sorts of cycles of enthusiasm and disillusionment that manifest themselves in communities such as Soperton, Waycross, or Columbus.

For our wood-based bioenergy research project, we initially selected our three primary research sites based on stages of bioenergy development: two liquid fuel plants in different stages of development (one mature but less successful and one developing but promising) and one operating wood pellet plant. For the first, we continued to work around the Range Fuels plant as it suspended operations and was sold to LanzaTech to produce aviation or other drop-in fuels. This research site represented a community that had gone through initial excitement and disappointment, but had continued hope, around bioenergy development. For our second liquid fuel plant, we first chose a promising proposed cellulosic ethanol plant, the Coskata plant in Boligee, Alabama, where the community was very engaged and enthusiastic about development. Shortly after our research began, Coskata terminated their Boligee development, and we substituted the nearby KiOR plant in Columbus, Mississippi. This plant was at the time the most advanced liquid fuels plant in the southern United States, and soon became the first plant to produce cellulosic fuel at a commercial scale. We completed our research around the KiOR plant just prior to its shutdown and bankruptcy in 2014. Its failure was likely due to technological difficulties and low gas prices—a result of the growth of fracking—which resulted in the plant being unable to produce cellulosic crude oil at competitive prices (Lane 2016). For our third research site, we chose the newly opened Georgia Biomass wood pellet plant in Waycross, Georgia, which was shipping wood pellets to Europe for power generation and was then the largest wood pellet plant in the world. We chose this plant because, unlike the two liquid fuel plants, it was in full operation and actively purchasing wood at a large scale (about 250 truckloads per day, or 1 million metric tons annually) (McCurry 2011), adding this dimension to our research. However, during our research, there were frequent rumors that the plant would close, and it was advertised for sale in June 2014 as RWE shifted its focus to other renewables.

The bioenergy plants we studied were sited based on proximity to what the forest industry refers to as "underutilized forest resources," as well as areas in need of rural development (Mayfield et al. 2007). While some smaller-scale bioenergy facilities use mill residues as feedstocks or are developed as bottom-up community initiatives, the three that we chose to focus on reflected efforts to develop a large-scale wood-based bioenergy industry in the southeastern United States based on direct harvests from forestlands. Our pilot research on new bioenergy plants suggested that the community and landscape effects of the two different types of bioenergy plants we studied were likely to be very similar because they involved the same raw material harvested in mostly the same way, the plants were at similar scales, and the long-term viability of both pellet and liquid fuel plants were often viewed skeptically. It is important to emphasize that our research was conducted during a period that combined enthusiasm and uncertainty for large-scale bioenergy development in the southern United States. Ultimately, development of commercial-scale wood-based cellulosic fuel plants in the southern United States has slowed dramatically, while export-oriented pellet plants have continued to operate under some uncertainty about the future of European renewable energy policies. These communities, and the bioenergy facilities located within them, are briefly described below.

Soperton, Georgia (Range Fuels/LanzaTech)

Construction began on Range Fuels in November 2007, after securing over $400 million in public and private funds. Range Fuels was expected to produce 40 million gallons per year of cellulosic ethanol using gasification technology and yellow pine as a feedstock - but only produced one batch of methanol. In rural and economically depressed Treutlen County, the initial announcement of the plant was met with great enthusiasm, as it would bring many jobs and a new market for wood products. As described earlier, the groundbreaking was attended by high-ranking government officials including Samuel Bodman, the U.S. secretary of energy. The local and national implications of Range Fuels' bankruptcy and closure in 2011 have been profound, leading to public anger over what is seen as a waste of taxpayer money. In 2012, New Zealand-based LanzaTech purchased the facility at auction for $5.1 million and renamed it the Freedom Pines Biorefinery. LanzaTech has retrofitted the facility for use as a research and development facility that has focused mainly on chemicals produced using proprietary microbes and synthetic biology (Schill 2014); in other plants, including in China, they have focused on production of sustainable aviation fuel (SAF) using gaseous by-products of steel production (including ATJ, or alcohol-to-jet platform technologies using captured pollution as a feedstock). As of November 2019, Michael Berube, acting deputy assistant secretary for transportation in the Office of Energy Efficiency and Renewable Energy of the DOE, stated that the DOE is currently negotiating with LanzaTech on a $14 million grant to create a demonstration-scale integrated biorefinery at the Freedom Pines site in Soperton, Georgia. He said:

> LanzaTech still has some remaining work to do under the initial award, and we have some negotiations to complete. But we're very excited about the prospects of this project and what it could mean for demonstrating the viability of drop-in biofuels in the United States. (Green Car Congress 2019)

While this particular use of the facility is not yet certain, our data, based on local experiences with Range Fuels, suggests that while some local people in Soperton (including landowners, community members, and local development authorities) will be cautiously optimistic that the facility will eventually employ more people, others are likely to express concerns about these new technologies being developed close to neighborhoods and schools.

Waycross, Georgia (Georgia Biomass)

Georgia Biomass, which began operation in 2011, has the capacity to produce 750,000 tons of pellets per year from local forests, which requires about 1.5 million metric tons of fresh wood per year (Gibson 2010). Georgia Biomass began as a wholly owned subsidiary of the German utility company RWE Innogy, and these pellets are shipped from the port in Savannah, Georgia, to supply biomass power plants and cogeneration facilities in Europe. Waycross, while more developed than Soperton, is also rural, with an economy heavily dependent on the forest products industry; the Georgia Biomass plant directly employed over eighty people and created over 300 indirect jobs (Argus 2014). In June 2014, the facility was offered for sale as RWE shifted its focus on renewables to more emphasis on solar and wind power (Statkraft 2015). Reports indicated that RWE sought a sale that would keep wood procurement around the facility and pellet supplies steady. However, in January 2015, RWE announced that it would not sell the plant (Statkraft 2015), and until early 2020, the Georgia Biomass facility continued to operate as a subsidiary of RWE. In early August 2020, Enviva Partners completed its acquisition of the Georgia Biomass facility, which will continue to operate as an industrial-scale wood pellet manufacturer. Extension agents from nearby counties said that while the plant has benefitted the communities, forest landowners have not yet made substantial changes to their forest management practices to supply it. Harvesting biomass for bioenergy would look similar to past harvesting for pulp mills, which have decreased in number. However, while the methods for growing biomass would not be unfamiliar to many forest landowners, the uncertainty of the new bioenergy market has led to reluctance for landowners to aim for biomass production instead of sawtimber and other products with more well-established markets. Some community members also expressed concerns about overharvesting of trees and impacts on air quality.

Columbus, Mississippi (KiOR)

After building a successful pilot plant in Pasadena, Texas, in 2010, KiOR built a demonstration facility and then the world's first commercial-scale cellulosic biocrude plant in Columbus, Mississippi, which began production in 2012 (figure 1.3). It used a proprietary biomass fluid catalytic cracking (BFCC) technique to convert biomass feedstock, specifically southern yellow pine, into crude oil that could be refined into gasoline, diesel, and aviation fuels. KiOR received a twenty-year no-interest $75 million loan from the state of Mississippi as an incentive to locate there, in addition to private investor funds. Promises by the company to provide over 1,000 jobs by the end of 2015 were not fulfilled, as the facility never reached full capacity and filed for bankruptcy in October 2013 (after our fieldwork there was completed). Following the chapter 11 bankruptcy, there have been a series of class-action lawsuits by shareholders, accusing the company of deliberately misleading them about chances of the company's success. In October 2015, the KiOR facility was sold to Georgia Renewable Power for $2.1 million, and it is likely that they will operate a small wood chip mill in that location (Smith and Hazzard 2015). It is important to note that Columbus, Mississippi, while still considered rural, has a much more developed and diversified economy

Figure 1.3   **KiOR Facility, 2014.** *Source*: Credit: Sarah Hitchner.

than our other field sites with several major industrial plants and a large military base.

CONCLUSION

Like all the big river systems of the southeastern United States, from the twisty Alabama to the relatively straight Chattahoochee, the history of wood-based bioenergy development in the region is full of springs and dams, convergences and divergences, and deltas and oxbows. It is full of stories of streams drying up and petering out and of levies breaking and flooding communities with elements both good and bad. Efforts to clean the rivers through public policies and citizen actions, or efforts by industries to merely green them through branding and publicity stunts (or to fight for the right to pollute them more) find parallels in the renewable energy story in the South. Like the rivers, and the trees that filter them and hold them in place, wood-based industries are the lifeblood of many rural communities in the region and have been for decades, even centuries. The promulgation of wood-based bioenergy facilities has promised to continue this tradition of dependency on forest resources, which has arguably led to the vast wooded tracts we see in this "woodbasket" region today. The story here about wood-based bioenergy development is one of many start-ups and failures and of both parallel and divergent trajectories.

The bioenergy story is also a play with many actors and almost as many stage directors, being enacted on multiple stages that are often not separate from the audience that watches them. Intense friction results from policies made by those who do not live here and thus do not see and feel their impacts. The story in one place is not the same as the story in any other. However, there are common threads running through these plotlines. Just as all water tends to run downhill, aiming for bigger rivers and ultimately the ocean if not absorbed or incorporated into some other end use, wood-based bioenergy development tends to follow the path of least resistance as it flows gently around obstacles when it lacks the power to simply barge through them with powerful waves and flow. In this book, we explore the place of wood-based bioenergy in the U.S. South and the landscape—social, cultural, histori-cal, political, and biophysical—in which these energy stories play out and the ways that various bioenergy development trajectories have both flowed around obstacles and flooded through them.

We are at a pivotal point in the history of not only this nation, but the world. Intensified racial and ethnic tensions, outmigration from rural areas, consolidation of agricultural and forestry landholdings, observable effects of climate change: there is a lot going on in our nation, and in the world, right now, and human actions today have a profound impact on the ability of future generations to meet their needs for food, shelter, safety, and energy. This belies the importance of bearing witness as a country, and we believe that ethnography can make real and substantive contributions to understanding the complexities within, and the connections between, these issues. There is a profound need for nuanced stories that portray the complex human and ecological dynamics in meeting current and future energy demands.

However, translating the reality of bioenergy development across multiple imaginaries and spheres of influence is a distinct challenge and forces difficult conversations about incommensurable ends when some groups stand to directly benefit from bioenergy development at the expense of others. Attention to the dynamics at play in the ecological, social, cultural, and political spheres is vital when assessing possible impacts of future bioenergy development trajectories on local communities. Multisited and multi-scalar ethnography adds depth to this conversation and backstory to these convoluted and overlapping narratives.

The twin processes of doing research and then writing about it have honed the broader questions that we seek to answer in our work. How, for example, can we learn to see beyond the surface of a landscape to the history it contains and the futures it might be part of? How can we as a nation facilitate honest and productive (if painful) discussions about race in the U.S. South to move past both explicitly racist events and the systemic racism and White privilege to which many non-minorities have been blind? How can we promote renewable energy policy and production in ways that are also beneficial to local communities that are directly affected by these developments? How can we talk about climate in ways that accept differences of opinion to move toward ecologically sustainable developments even without consensus? In examining the ways that energy, and especially wood-based bioenergy, intersects with landscapes, climate, and race in the U.S. South, we do not aim to provide neat answers to these questions. Rather, we believe that telling the stories of recent bioenergy developments in the region and the perceptions among local community members of their successes and failures in living up to their promises and potential is one way to measure where we are as a nation in terms of environmental sustainability, racial equality, and concern for future generations.

NOTES

1. Name changed for privacy.
2. After consulting numerous authoritative sources and deliberating extensively, we have chosen to use the capitalized forms of "Black," "White," "Brown," Indigenous," and "Tribal" when referring to racial and ethnic identity throughout

this book. There is no consensus regarding capitalization, and standards within the academic and popular publishing realms are in a state of flux. However, we feel that using this language is the most equitable and inclusive way to acknowledge the historical and social construction of racial and ethnic identity and the very real consequences it has for all people in our society.

3. See: https://www.dogwoodalliance.org/our-work/our-forests-arent-fuel/

4. FERN, BirdLife Europe/Royal Society for the Protection of Birds [RSPB], ClientEarth, European Environmental Bureau, Greenpeace

5. FERN, Suomen Iuonnonsuojeluliitto [Finnish Association for Nature Conservation], Forests Monitor, The Woodland League, Friends of the Earth Europe

6. "The Agriculture and Food Research Initiative (AFRI) is the nation's leading competitive grants program for agricultural sciences. The National Institute of Food and Agriculture (NIFA) awards AFRI research, education, and extension grants to improve rural economies, increase food production, stimulate the bioeconomy, mitigate impacts of climate variability, address water availability issues, ensure food safety and security, enhance human nutrition, and train the next generation of the agricultural workforce." (https://nifa.usda.gov/program/agriculture-and-food-research -initiative-afri)

7. Other ethnographic work on these topics was supported by the Southern Research Station, USDA Forest Service ("Social and Cultural Dimensions of Climate Change Adaptation: Anthropological and Sociological Approaches to Social Vulnerability and Biofuels in the U.S. South," 2011–2016) and USDA National Institute of Food and Agriculture-Agriculture and Food Research Initiative (NIFA-AFRI) Sustainable Bioenergy Challenge Area ("Social Acceptability of Bioenergy in the U.S. South," 2012–2016).

8. For example, Forest Bioenergy Conference in Forsyth, Georgia in 2011; Wood Bioenergy Symposium in Atlanta, Georgia in 2012; SunGrant Initiative National Conference in New Orleans, Louisiana in 2012; and Mississippi Biomass and Renewable Energy Council Annual Conference in Tunica, Mississippi in 2013.

Chapter 2

What People Hear and What People Say about Bioenergy

Translating Bioenergy Narratives, Imaginaries, and Metaphors

BIOENERGY: IS THAT A PROMISE OR A THREAT?

Proponents of bioenergy paint a pretty picture. They claim it simultaneously promises to reduce dependence on fossil fuels (de Gorter and Just 2009; Marland et al. 1997), contribute to global and national energy security (Evans and Cohen 2009), produce a carbon negative or neutral fuel source and, therefore, mitigate climate change (Cook and Beyea 2000; Johnson et al. 2007; Hill et al. 2009; Perlack et al. 2005), and serve as a vehicle for rural development (IEA 2002a, 2002b; Sims 2003). However, this is a simple narrative, one that fails to acknowledge a number of complex trade-offs, the existence of contested data, or incommensurable assumptions. Numerous counternarratives question these assumptions. Some of these sets of narratives and counternarratives include: (1) claims that growing biomass for energy uses arable land that should be used to grow food versus the argument that biomass-based fuels are made from agricultural wastes or sources that otherwise do not displace food production ("the food versus fuel debate"), (2) the claim that biofuels lead to deforestation and the alternate possibility that energy crops use forest industry by-products or can even contribute to ecological restoration, (3) the debate over whether biofuels are—or are not—in fact carbon and greenhouse gas negative or neutral once a complete life cycle analysis is conducted, (4) claims that biofuels can bring rural development and critiques of the kinds of development it brings, and (5) the claim that biofuels development spurs technological innovations, tempered with the fears of what the unintended consequences these new technologies may be.

As we explore in this chapter and in more detail in the following chapter, it is also important to remember that in rural, forest-dependent communities, landowners' willingness to manage their forests specifically to provide feedstock for biofuel markets based on purely economic calculations cannot be taken as a given (Delshad et al. 2010; Fast 2009; Petrou and Pappis 2009; Plate, Monroe, and Oxarart 2010; Selfa et al. 2010; Sengers, Raven, and Venrooij 2010; Susaeta 2010), however much economists would like to believe that. As we explore later, landowners, both forest landowners and farmers, vary in their adherence to risk aversion strategies (Caldas et al. 2014); many are wary of "putting all their eggs in one basket," especially when that basket is an unfamiliar entity.

Members of the general public, when they think about bioenergy at all, are influenced by a number of sources that have varying levels of credibility. They hear messages from different actors operating on different scales both directly and filtered through various media sources, and they also circulate these messages among themselves. People's experiences with bioenergy facilities built in their hometowns, including the positive and negative impacts they have had on different members of the community, directly affect public perceptions of bioenergy. Understanding what people hear about bioenergy, and from what sources, is especially important because local support is often critical for both new industries and for local governments seeking to lure them to rural communities that are often economically depressed. To achieve this understanding, it is necessary to evaluate the ways that different actors are talking about bioenergy in different contexts, their goals and motives in discussing bioenergy, the narratives and rhetorical devices that they use to strategically influence people or pass on ideas and opinions to others, and the dynamics of translation across different scales and imaginaries. A number of scholars have discussed the role that discourses, narratives, metaphors, and imaginaries play in shaping the opinions of various stakeholders in energy politics (including international policy-makers, bioenergy industry representatives, members of local communities affected by large or small-scale energy projects, and the general public), and several studies are specific to bioenergy development (Curran 2012; Gasparatos et al. 2011; Kirkels et al. 2012; Kuchler and Linnér 2012; Levidow and Papaioannou 2013; Miller et al. 2014; Omer 2013; Spartz, Rickenbach, and Shaw 2015; Trutnevyte 2014; Vel 2014). We build on this body of work while paying particular attention to translational dynamics.

In this chapter, we focus on the ways that bioenergy imaginaries are articulated, responded to, and strategically altered and interpreted by people who are drawing on diverse experiences, values, and interests. First, we ground this discussion in a brief description of the various types of woody biomass products that are produced in the southeastern United States, provide an overview of wood-based bioenergy policy, and offer an explanation of how various stakeholders in this region frame issues related to bioenergy. Second, we discuss imaginaries and address the main elements that make up the pieces of several prominent bioenergy imaginaries, which emerged as themes during our ethnographic research: (1) energy security, (2) rural development, (3) government promotion and investment, (4) climate change, (5) forests and forest products, and (6) renewability and sustainability. We will discuss these in more detail in the following chapters but present them here briefly in the context of bioenergy narratives and imaginaries. Third, we discuss which bioenergy narratives and metaphors various actors employ, and how recurring linguistic elements are shared among bioenergy stakeholders. We focus on several key metaphors that people reference when they talk about bioenergy in different contexts, including public media, policy and management discussions, bioenergy conferences, outreach programs, and among landowners and within communities: (1) "snake oil," (2) "silver buckshot," and (3) "people who hate us" (i.e., terrorists). Fourth, we explain how these metaphors employ multiple, overlapping, and sometimes conflicting or incommensurable conventional discourses (Strauss 2012) in order to appeal to emotions and cultural value systems, and we argue that uses of these metaphors act as "moments of influence" (Witter et al. 2015) on perceptions of bioenergy. Fifth, we link this discussion of linguistic analysis of the words that people use then talking about bioenergy back to the concept of imaginaries, showing how the elements of various imaginaries are interwoven with broader cultural values and visions for the future (Hitchner, Schelhas, and Brosius 2016). We highlight the ways that various actors use specific language both to pass on ideas unintentionally and to strategically use bioenergy imaginaries to advance certain interests.

ENERGY AND ENERGY POLICY

Even defining bioenergy is problematic, given the difficulty in separating the physical energy products from the discourses in which they are embedded. Kuchler and Linnér (2012, 582) state that "bioenergy is perceived as a conceptual entity formed and advanced as a carrier of specific meanings for political and economic reasons." Vel (2014, 2816) applied the phrase "discursive commodities" (from Fairhead, Leach, and Scoones 2012) to Jatropha as a biofuel feedstock. She defines discursive commodities as "objects of trade that have obtained market value because of the narratives that science, technology, politics, and business have created about them, but do not exist yet in the real world." Like Jatropha, and other bioenergy products and technologies that do not yet exist or are not yet operable on a commercial scale, they acquire value when they reference discourses such as green development, rural employment, land tenure security, and renewable energy. Portz (2014) notes that pine in the southeastern United States is a sustainable tree crop that has multiple end-uses and guaranteed markets, including pellets, for which there is an established market. Pine, therefore, may be considered a "conceptual entity," or "discursive commodity," only when it is transformed into a hypothetical product—or one not yet produced at a commercial scale—such as cellulosic biocrude. We found that while many community members and forest landowners across the region did not have a sophisticated understanding of the various bioenergy feedstocks and technologies, both current and potential, they do have a deep, and deeply nuanced, understanding of forests as places of economic as well as cultural, or even spiritual, value. We also found that even among forest management professionals and wood-based bioenergy industry insiders, forests are more than simply utilitarian.

As the southeastern United States has a warm climate with mostly sufficient and predictable rainfall, it is conducive to the high vegetative productivity of many leading bioenergy crops. It also has well-developed extant forestry and agricultural industries. Because of these two main factors, there has been, and will likely continue to be, a rapid expansion of the biofuel industry in this region as the demand for biofuels increases both domestically and internationally (Evans and Cohen 2009; Kim and Hayes 2006). There is a wide and still-unfolding array of energy technologies and feedstocks here which includes: (1) pellets from woody biomass and energy grasses; (2) biodiesel from oilseed crops (such as soybean, oilseed rape, crambe, and camelina); algae, and animal fats; and (3) ethanol from traditional food crops such as corn, sugarcane/bagasse, wheat, and rice (Aryee et al. 2011; Baker and Zahniser 2006; Botha and von Blottnitz 2006; Carlson et al. 1996; Demirbas 2006, 2011; di Giacomo and Taglieri 2009; di Pardo 2000; Dias de Oliveira et al. 2005; Dwivedi et al. 2011; Hill et al. 2006; Marris 2006; Nelson 2002; Pate, Klise, and Wu 2011; Perlack et al. 2005; Putnam et al. 1993; Swaan and Melin 2008).

A variety of technologies and feedstocks are also involved in the production of cellulosic ethanol from nonfood sources. Feedstocks can include dedicated bioenergy crops such as trees (Kszos et al. 2001; Lemus and Lal 2005; Stanturf et al. 2001; Perlack et al. 2005) and perennial energy grasses (Sanderson et al. 1996; McLaughlin et al. 2002) such as alfalfa (González-García 2010), switchgrass (Vogel 1996, 2004), and Miscanthus (Jørgensen 2011). A variety of agricultural residues and by-products (including corn stover and livestock manure), forest industry by-products (including harvest residues and sawdust), and municipal waste (Dickman 2006; FAO 2004; Graham et al. 2007; Hoskinson et al. 2007; Perlack et al. 2005; Smith et al. 2004) can also be used as feedstocks. Bioenergy and biofuel products that use wood as a feedstock include wood pellets and liquid fuels such as cellulosic ethanol and biocrude. Wood pellets are made from wood fibers extracted from whole logs, parts of logs, or what is considered "wood waste" or unmerchantable parts of trees such as tops, limbs, or leaves/needles. These are burned directly or co-fired with other materials such as coal or even tires (making tire-derived fuel, or TDF), producing heat and/or electricity. Cellulosic ethanol is made from pine trees, and major technologies to produce cellulosic ethanol generally follow one of two pathways: (1) a series of processes to transform cellulosic materials to sugars such as glucose and xylose using steps involving pretreatment, enzymatic hydrolysis, and fermentation (Sheehan et al. 2002; US DOE 2006) or (2) thermochemical processes such as co-firing, gasification, and various methods of pyrolysis that produce heat and energy through combustion (Demirbas and Arin 2002; Yaman 2004). Cellulosic ethanol can be blended into traditional petroleum-based gasoline, while biocrude is a drop-in liquid fuel that replaces fossil fuels using existing infrastructure. KiOR was the first commercial-scale cellulosic biocrude producer, using next-generation catalysts and a proprietary biomass fluid catalytic cracking, or BFCC, technology (Lane 2012). As noted, they were ultimately unsuccessful. In any case, while each of these different technology applications uses trees—in the case of the U.S. South, mostly southern yellow pine—as primary bioenergy feedstocks, they are in different stages of experimentation and large-scale production and thus have different real and perceived effects on the landscape. They also create different sets of opportunities and outcomes for multiple stakeholders, including community members and forest landowners near newly built bioenergy facilities.

While broad global, national, and regional benefits are important, specific bioenergy developments will entail real changes in local landscapes and complex value trade-offs at different levels. Biofuels are not just about energy; they connect many different issues and realms of governance, including agriculture and food, forest management, international trade relations, global climate change, landscape conservation and ecosystem services, and a host of social, cultural, and economic issues. Sims (2003) notes that many energy-producing companies, previously focused on their "bottom line," are now examining their "triple bottom line" of ecological, economic, and social impacts on society.

The Triple Bottom Line

Ecological Impacts

Ecologically, demand for biofuels has led to deforestation and landscape fragmentation in many parts of the world (Koh and Wilcove 2008; Laurance 2007; Nepstad et al. 2001); there is also concern that production of biofuels can increase net greenhouse gas emissions rather than decrease them (Gibbs et al. 2008). It has also led to the conversion of forested, agricultural, and pasture lands to intensively managed monocrop plantations (Fargione et al. 2008; Righelato and Spracklen 2007; Searchinger et al. 2008), with a subsequent loss of agrobiodiversity (Fletcher et al. 2011) and soil fertility (Huggins et al. 1998; Johnson et al. 2006; Vitosh et al. 1997) and an increase in soil erosion (Graham et al. 2007; Wilhelm et al. 2004). Many bioenergy crops have heavy water needs (Evans and Cohen 2009; Giampietro et al. 1997; NRC 2007) and require large inputs of fossil fuels, in the form of fertilizers, as well as for harvesting, transportation, and production (Mann et al. 2002; Perlack 1992; Shapouri 1995; Wilhelm et al. 2004). Many bioenergy crops, which may be exotic or genetically modified, also have the potential to become invasive (Barney and di Tomaso 2011; Linacre and Ades 2004; Wilkinson and Tepfer 2009). However, genetic modification of energy crop plants can also reduce water needs and crop damage from pests and diseases and increase crop yields (Chunxiang 2011; Tammisola 2010). Also, in some areas, bioenergy crops have been grown on formerly degraded or marginal lands, possibly contributing to ecological restoration efforts by increasing wildlife habitat, reconnecting fragmented landscapes with vegetation, intercropping bioenergy-specific crops with traditional crops, and increasing carbon sequestration in soils (Beyea 1994; Christian 1998; Hill et al. 2006; Lee et al. 2007; McLaughlin and Walsh 1998; Sanderson et al. 1996).

Researchers and government agencies have conducted studies assessing the sustainability of biofuel production in the United States (Buford and Neary 2010; Damery et al. 2009; Janowiak and Webster 2010; Richardson et al. 2002; Perlack et al. 2005; US DOE 2011), many using common parameters for life cycle analyses of biofuels such as ethanol, including agricultural inputs (seed, labor, pesticides, herbicides, irrigation, machinery, labor), transportation inputs (fuels and machinery), feedstock outputs and yields, ethanol conversion inputs (fossil fuels, chemicals, and machinery), and ethanol yields (Giampietro et al. 1997; Gnansounou et al. 2005; Macedo 2008; Schmer et al. 2008; Shapouri et al. 2002; Wang et al. 2007). Evans and Cohen (2009, 2261) compared life cycle analyses for four dedicated energy crops (sugarcane, sweet sorghum, corn, and southern pine) in Georgia and Florida in terms of their energy balances and water requirements; each crop has pros and cons, and they conclude that "careful scrutiny of environmental trade-offs is necessary before embracing aggressive ethanol production mandates." Government agencies and researchers have been actively working to find ways to mitigate the detrimental effects of large-scale bioenergy crop production and to encourage the conversion of eroded, overgrazed, degraded, and unused farm and pasture lands to sustainable farming and forestry practices that can supply feedstocks for bioenergy facilities (Lattimore et al. 2009; Lora et al. 2009; Marshall 2007; Reijinders 2006).

Economic Impacts

Biofuels can potentially encourage rural development, job creation, and diversification of income for landowners and local communities (IEA 2001, 2002a, 2002b; Sims 2003), although some researchers have indicated concerns regarding a lack of skilled and semi-skilled workers that are willing and able to fill the newly created positions (Sims 2003) and the low-wage and potentially high-risk jobs that may be created (Gunderson 2008; Rossi and Lambrou 2008). The United States has in place many subsidies, tax credits, and other financial incentives for farmers and biofuels producers (de Gorter and Just 2009; Duffield and Collins 2006; EESI 2006; Gielecki et al. 2005; Guo et al. 2007; Hill et al. 2006; Miranowski 2007; Nazzaro 2005). Biofuels produced from food crops such as corn have affected food prices and supplies of food (Mitchell 2008; Naylor et al. 2007); public concerns over both the feasibility and the ethics of converting arable land to biofuel croplands have led the drive to develop biofuel feedstocks that do not compete with food supplies (Evans and Cohen 2009; Hill et al. 2006; Kim and Hayes 2006; Pimental and Patzek 2005). As we review in the next chapter, making this economically feasible using woody biomass, even in heavily forested areas, is an ongoing challenge, especially when oil and gas prices are low. In addition, the economic impacts of bioenergy development in rural, forest-dependent communities are not uniform; within communities, there are those who benefit more than others, while others pay the price in both monetary and nonmonetary terms.

Social Impacts

There are also many important social and cultural impacts associated with biofuels, some specific to the southeastern United States, and it is critical to address these issues as bioenergy industries develop in this region. Landscape history and traditional land values are in some areas threatened with dramatic land use changes associated with large-scale bioenergy crop production; these can in turn have effects on social cohesion and stability. Forests in the South are important to people for timber, wildlife, biodiversity, watershed protection, recreation, and legacy values (Wear and Greis 2002). However, development in the biofuels industry in the United States can also be an important vehicle for rural development, community revitalization, and stemming the flow of rural-to-urban migration (Sims 2003). There are also complex interactions between areas that supply raw materials for biofuels and areas that import them; this has led to scrutiny of extractive economies for their inequitable distribution of the costs and benefits of bioenergy development (Hornborg et al. 2007). For example, much of the demand for biofuels occurs in the Global North, while the majority of biofuels are produced in the Global South. Similar dynamics are playing out within the Global North as well, as much of the current pellet production in the southeastern United States supplies European markets (Dwivedi et al. 2011; Sikkema et al. 2010).

Bioenergy development has different effects on various populations within the southeastern region as well, from individual landowners and local communities to industry leaders and policy-makers. In natural resource management and economic development, racial and ethnic minorities have often borne a disproportionate share of the impact or received few opportunities (Schelhas 2002). It is not uncommon for the individual interests of minority stakeholders to be neglected when pursuing broad goals such as forest conservation or energy independence that are defined as being in the public interest (Schelhas and Pfeffer 2008). Bullard (2011) has cautioned that " 'green' biomass (like energy crops)" can have environmental justice impacts and expose poor and minority communities to more toxins from fertilizer application in bioenergy crop plantations to pollution of air and water surrounding biofuel production facilities (Hill et al. 2009).

In addition to health concerns, Sims (2003, 365) notes other "social barriers" that

> need to be overcome in order to obtain local support and resource and planning consents, particularly where a waste-to-energy plant is planned, often being subject more to emotional judgment in the media rather than to any scientifically-based reaction from a well-informed public.

People's perceptions of the ecological and economic impacts of biofuels are colored by their experiences, worldviews, value systems, and access to knowledge and resources. There is a great deal of complexity regarding bioenergy development, and while there is much technological and economic complexity and uncertainty, much of the inherent complexity is social in nature.

The development of the liquid biofuels industry in the U.S. South is in the early stages and has seen some fluctuations, and most landowners and community members are now cautiously optimistic about a bioenergy industry and related markets. There is limited investment due to uncertainty, careful attention to long-term job creation and development potential, and interest in environmental sustainability. In some sites, there has been opposition to bioenergy development. Interest in biofuels is usually justified on the basis of energy independence and rural development. However, the social acceptability of bioenergy is an ongoing concern among bioenergy industries, especially given the prevalence of anti-ethanol sentiment, especially in the U.S. South (see figure 2.1 for a photo of one sign, of many, that we have seen in the region) (Brosius, Schelhas, and Hitchner 2013; Hitchner et al. 2014; Hitchner and Schelhas 2012; Hitchner, Schelhas, and Brosius 2017; Schelhas,

Figure 2.1 **Anti-ethanol Sign at Gas Station.** *Source*: Credit: Sarah Hitchner.

Hitchner, and Brosius 2018). Our interviews with diverse stakeholders suggest that no one has a complete understanding of what is happening at any particular site, and we hope that this book, which focuses on the social impacts of bioenergy development, can play a role in bringing clarity to the diverse perspectives and interests.

GOVERNANCE OF BIOENERGY

Special interest groups, such as environmental NGOs (i.e., Dogwood Alliance, Sierra Club), professional bioenergy associations (i.e., American Bioenergy Council [ABC], American Council on Renewable Energy [ACORE], National Biodiesel Board [NBB]), or international, multi-stakeholder groups that aim to influence bioenergy policy (i.e., World Bioenergy Association [WBA], International Renewable Energy Association [IRENA], Roundtable on Sustainable Biomaterials [RSB]) operate at different scales (community, regional, national, international). Numerous studies regarding governance of biofuels have been conducted at different spatial and temporal scales: global (Hitchner 2010; Maclin and Dammert Bello 2010; van der Horst and Vermeylen 2011; Scott et al. 2014), national (in this case, the United States) (Bourgeon and Trueger 2010; Gillon 2010; Bailis and Baka 2011), and regional (e.g., the southeastern United States) (Evans and Cohen 2009; Abt et al. 2010; Bailey et al. 2011).

Policy-led efforts to develop a renewable energy system and promote rural development through wood-based bioenergy development can be usefully analyzed through the concept of sociotechnical imaginaries. In the United States, energy imaginaries entailing energy security and energy independence as an alternative to fossil fuels have long been part of the rhetoric of politicians, though most have to carefully toe lines between groups of constituents. This narrative promoting energy independence, which crosses party lines, dates back to the 1960s and 1970s (Bailey, Dyer, and Teeter 2011; Strauss, Rupp, and Love 2013; Tidwell and Smith 2015) but intensified in the United States after the terrorist attacks of September 11, 2001. In 2006, George W. Bush lamented the U.S. "addiction to oil," while in 2007, Barack Obama promoted freedom from the "tyranny of oil" (Bryce 2008). This type of rhetoric evokes emotional reactions in citizens in support of alternate sources of energy and merges with environmental discourses about renewable energy-reducing emissions and mitigating climate change.

Notably, former president Donald Trump promised "complete energy independence" in 2017, although instead of replacing fossil fuels, he aimed for "American energy dominance" which he hoped to achieve by exploiting the country's "vast untapped domestic energy reserves" by removing "burdensome regulations on our energy industry" (as recorded by then-interior secretary Ryan Zinke in Trump's Secretarial Order 335: Strengthening the Department of the Interior's Energy Portfolio). Trump was particularly vocal about the ways that biofuels, especially ethanol, pit powerful oil and gas interests against agribusiness and farmers. While trying to craft biofuel quotas and other regulatory mechanisms that would appease both farmers and oil industries, he approved exemptions for small oil refineries from the requirements of renewable alternatives, which led to a wave of biorefinery closures, further hurting farmers already suffering from trade tensions with China. *Bloomberg* (Jacobs et al. 2019) reported on September 19, 2019, that Trump tried to navigate these political pressures with little success: "President Donald Trump on Thursday expressed his rising frustration over trying to reach an accord between warring oil and biofuel interests by saying the negotiations were more difficult than dealing with the Taliban."

Current president Joe Biden, immediately after inauguration, rejoined the Paris Climate Accord and began to pursue environmental regulations that will cap methane pollution limits by oil and gas corporations, introduce rigorous new fuel economy standards, require light- and medium-duty vehicles to emit zero emissions, and ban new leases on oil and gas exploration on publicly owned land. He has also pledged to earmark $400 billion over the next ten years for clean energy development. Biden's promotional webpage states that he "will ensure our agricultural sector is the first in the world to achieve net-zero emissions, and that our farmers earn income as we meet this milestone" (www.joebiden.com). Given that Trump's relationship with biofuel advocates was contentious (mainly due to providing more subsidies and drilling and fracking permits for fossil fuel producers, as well as exemptions to biofuel blending mandates for transportation fuels, making it more difficult for clean energy companies to compete), Biden worked during his transition period before taking office to meet with biofuel companies and trade groups to listen to their concerns and reiterate his support for them (Kelly 2020). Much of this support will come through policy designed to enforce higher blending walls,[1] or percentages of renewable fuels mixed into petroleum-based transportation fuels. He is planning to roll back waivers for blend requirements enacted by the Trump administration (Dloughy 2021), and he has pledged support for further development of cellulosic biofuels as well.

While potential changes to these policies are currently being negotiated within the Biden administration, currently in the United States, the U.S. Energy Policy Act [EPACT] of 2005 and the Energy Independence and Security Act [EISA] of 2007 mandate the use of renewable energy sources. EPACT established a mandatory Renewable Fuel Standard (RFS), which was expanded by EISA. EISA mandates a six-fold increase of ethanol usage in the United States by 2022 (to 36 billion gallons a year, of which only 15 billion gallons can be corn ethanol; advanced biofuels, including cellulosic ethanol, account for the other 21 billion gallons) (Halvorsen et al. 2009). However, the United States lacks a national bioenergy policy, and states have individual mandates and incentives for bioenergy development. For example, Wisconsin has a goal of 25% renewable transportation by 2025 (Halvorsen et al. 2009). Michigan offers E85 infrastructure development grants, fuel tax reduction for high blend fuels, and renewable energy tax incentives (Halvorsen et al. 2009), while the Missouri Qualified Fuel Ethanol Producer Incentive subsidizes ethanol made from Missouri agricultural products (Aguilar and Garret 2009).

While the feasibility of large-scale bioenergy production and its potential to displace conventional fossil fuels have long been questioned, the U.S. Department of Energy's "Billion Ton" reports (Perlack et al. 2005; Perlack and Stokes 2011; Langholtz, Stokes, and Eaton 2016), which examined the feasibility of an annual supply of one billion tons of biomass as a feedstock for bioenergy, determined that the national policy goals for renewable energy production were achievable. Together, these goals link the need for energy independence and security with rural development. Nongovernmental organizations also joined the effort; one example is 25×'25, which defines itself as "a diverse alliance of agricultural, forestry, environmental, conservation and other organizations that are working collaboratively to advance the goal of securing 25 percent of the nation's energy needs from renewable resources by the year 2025" (25×'25 2010). The forest resources in the southeastern United States continue to be seen by various national and regional entities as integral to achieving energy independence and security, developing a national renewable energy portfolio to address climate change, and stimulating rural development.

Bioenergy governance in the European Union, which has been intensively developed since early 2000s (Rohracher et al. 2004), includes mandates for renewable energy that directly affect production of wood-based bioenergy products (especially wood pellets) in North America and around the world (di Lucia 2010; Sharmon and Holmes 2010; Upham et al. 2011). Several studies have noted concerns among members of the general public in Finland (Pätäri 2010), the United States (Selfa et al. 2011), and other countries that the forest bioenergy business is dependent on policy and that prospects could change quickly with policy change; the unpredictability of policy, particularly in the European Union, is a factor heavily influencing public opinion about the long-term viability of forest-based bioenergy. Also, Söderberg and Eckerberg (2013) note that E.U. bioenergy policy potentially conflicts with E.U. targets for biodiversity conservation and land use change, which further complicates public perceptions of the ecological sustainability of forest-based bioenergy.

Certification is a form of market-based governance that is being incorporated into bioenergy development in order to ensure renewability and sustainability of energy feedstocks. There are numerous certification systems for forest and forest products in place around the world; these include (but are not limited to) SFI (Sustainable Forestry Initiative), FSC (Forest Stewardship Council), American Tree Farm, SmartWood (Rainforest Alliance), PEFC (Programme for the Endorsement of Forest Certification), and the Green Gold Label (Woodwork Institute 2004; Newsom and Hewitt 2005; Newsom et al. 2006; Overdevest and Rickenbach 2006; Van Dam et al. 2008; Ginther 2011; Gan and Cashore 2013). Requirements or incentives for certification will influence landowners' forest and agriculture management decisions, as well as public perceptions of wood-based bioenergy. Monroe et al. (2011, 82) note that "One way to mitigate concerns about forest sustainability may be through forest certification systems that establish criteria for sustainable forest management." However, others cite concerns about the certification of sustainable biofuels because of extra cost and burden to producers and landowners (Selfa et al. 2011). As noted by Magar (2011), there appears to be a tension: while bioenergy experts tend to see certification as a solution, entrepreneurs or landowners may see it as a problem. For the general public, though, certification could be a way to ensure sustainability and thus mitigate possible concerns and improve the general social acceptability of wood-based bioenergy.

From enforceable, policy-based regulatory mechanisms down to individual landowner incentive programs, bioenergy development is driven by various visions of the role that bioenergy can play in meeting renewable energy standards or diverting funding from more promising alternatives, contributing to rural development or exacerbating extant socioeconomic pressures, mitigating climate change or perhaps making it worse, and preserving forests by supporting forest-dependent communities or in overexploiting forest resources in a neocolonial fashion. Each of these visions of bioenergy offers a seemingly holistic, but in reality, partial, perception of how bioenergy can affect land-scapes, landowners, and communities on different scales. Next, we examine how these visions about bioenergy can be considered as imaginaries, and how the elements of which they are composed are neither static nor fixed in place.

IMAGINARIES

The past several decades have seen a proliferation of research within a field of knowledge called envirotechnical analysis or envirotech research, which includes study of the cultural and political histories of technology (Latour 1993; Haraway 1991; Sorenson 1991; Stine and Tarr 1998; Pritchard 2011, 2013; Russell et al. 2011; Turner 2015; Mittlefehldt 2018; Jeon and Kang 2019; Macfarlane 2020). Sara Pritchard explains envirotechnical regimes as "institutions, people, ideologies, technologies, and landscapes that together define, justify, build, and maintain a particular envirotechnical system as normative." We contend that both actual and discursive bioenergy commodi-ties, from wood pellets to cellulosic biocrude, are emerging as key elements of these envirotechnical regimes and that they are embedded in what theo-rists of critical discourse analysis (CDA) and science and technology studies (STS) call "imaginaries," as presented in the last chapter. As noted, Jasanoff and Kim (2009, 2013) introduced the notion of "sociotechnical imaginaries," which builds on a broader literature on social imaginaries (Anderson 1983; Castoriadis 1987; Taylor 2004) and connects these to technological develop-ment (Jasanoff and Kim 2009, 2015). Sociotechnical imaginaries shape social transitions to new forms of energy, including renewables sourced from local resources. These imaginaries are driven by various narratives and discourses, and Levidow and Papaioannou (2013, 36) argue that bioenergy innovations are a critical part of various sociotechnical imaginaries in which fuels and power sources are produced in an ecologically sustainable and socially just manner. Genus et al. (2021, 2) say that "More directly than narratives, socio-technical imaginaries serve an 'explanatory and justificatory' purpose, though they are considered to be less explicit and accountable than policy agendas." Instead, they promote encouraging visions of what the future can hold.

Imaginaries have material outcomes in that they guide government policies and investment in science and technology toward a future energy system. Fairclough (2010, 480) states that "Imaginaries produced in discourse are an integral part of strategies, and if strategies are successful, then associated imaginaries can become operationalized, transformed into practice, made real." Genus et al. (2021, 8) note that "To invoke energy imaginaries is to look to the future in such a way as to govern contemporary practice . . .

Imaginaries are materialised as individuals talk and act with others about energy." Through the sharing of ideas is how these visions become real.

Visions and the policies built upon them influence trajectories of technological development, channel public expenditures, and shape the inclusion or exclusion of citizens in the benefits and costs of technological development (Jasanoff and Kim 2009). They are also social, cultural, and moral in that they are imbued with visions of what is good or desirable for society and that they result in actions that reconfigure patterns of life and work, as well as the allocation of benefits and burdens from energy systems (Jasanoff and Kim 2009, 2013, 2015). The dominant techno-economic energy imaginary foregrounds the imperative to find the most efficient and cost-effective solutions to meet the goal of ensuring energy security with reduced carbon emissions. This approach emphasizes the need for technological advances and the social acceptability of renewable energy technologies and behavioral changes by energy producers and consumers (Genus et al. 2021, 3). Genus et al. (2021) present an alternative energy imaginary in which risks of technological failure are factored in and bottom-up governance of energy systems supplant top-down techno-economic approaches. While sociotechnical imaginaries are generally seen as emanating from states, they are not specific policy agendas but rather shared understandings that set desirable goals, articulate feasible futures, and create political will and public resolve to attain these visions (Jasanoff and Kim 2009). A growing body of research shows that while local and everyday talk may not directly influence dominant energy imaginaries, which place technological innovation at their center, they do affect how people on the ground experience and communicate about them (Genus et al. 2021; see also Levidow and Papaioannou 2013; Engels and Münch 2015; Ballo 2015; Delina 2018; Kuchler 2014; Tozer and Klenk 2018; Schelhas, Hitchner, and Brosius 2018).

Imaginaries are considered "group achievements" and are collectively held (Jasanoff 2015), but recent research has shown that they can be contested (Kim 2015) and can emerge from below the "seats of power" (Jasanoff 2009, 20). Imaginaries are increasingly viewed as dynamic, unfolding historically through individual actions, shaped by powerful actors but requiring support from coalitions, and subject to contestation and replacement by alternatives (Kim 2015; Hurlbut 2015; Jasanoff 2015). However, despite the holism of bioenergy imaginaries, most research studies focus on specific aspects of bioenergy production and distribution, and there is ongoing debate about the applicability of studies about one region, feedstock, or production pathway to others. The effects of bioenergy development on local places are complicated and uneven. Eaton et al. (2014) note that "attempts to achieve specific imaginaries through technological projects become sites of contest and conflict," especially in local communities in which bioenergy facilities are located.

The dynamics of local resistance or acceptance of bioenergy are difficult to pinpoint (Devine-Wright 2007), though an analysis of sociotechnical imaginaries can help to fill in the gaps of this understanding (Eaton, Gasteyer, and Busch 2014) while drawing our attention to the translational dimensions of bioenergy development in particular places and across scales. Further, Vel (2014) argues that global discourses are framed on an international level, while localized discourses operate where feedstocks are actually produced in specific areas. Genus et al. (2021, 2–3) note that

> much research has focused on state energy imaginaries and expert-driven views thereof, which tends to neglect the "underlying messiness" of, and variation in, energy imaginaries; this messiness and variation can be grasped more effectively by studying "multivalent" stories in local places, though studies of national energy imaginaries may highlight what visions of the public (s) are being employed and acted upon by policy-makers, technical experts and researchers.

With this in mind, here we try to address the social complexities associated with wood-based bioenergy in the U.S. South from multiple angles, elucidating the prioritized elements of imaginaries emanating from various actors who operate on different scales. There are complicated, sometimes conflicting, dynamics between actors making decisions based on these multiple discourses, and in these dynamics, the multi-scalar nature of discourses becomes clear (Scott et al. 2014; Kuchler and Linnér 2012), as we shall see in the following chapters.

Recent research on energy imaginaries has explored their complexity, reinterpretation, and contestation (Delina 2018; Tidwell and Smith 2015; Burnham et al. 2017; Smith and Tidwell 2016). There are also several recent publications in the subgenre of energy ethnography (Franquesa 2018; Watts 2018; Whitington 2019), which examine myriad effects of the development of various energy, including alternative energy, initiatives on local communities. Several other books (i.e., Strauss, Rupp, and Love 2013; Imre and Boyer 2017) explore the contributions of an anthropological or humanistic approach to energy studies, and there are other publications that explore links between racial inequalities, climate change, social justice, neoliberal economics, technological innovations, and forested landscapes (i.e., Klein 2014; Taylor 2014; Jasanoff and Kim 2015). However, to our knowledge, there is no other multisited energy ethnography set in the southeastern United States, known as the "woodbasket" of the country and home to a number of economically depressed forest-dependent communities, that explores wood-based bioenergy in the framework that we offer which uses ethnographic examples to weave together complicated and difficult discussions about energy, forests, climate, and race through the lenses of imaginaries, scale, and translation. Bringing attention to these imaginaries provides the context necessary to understand the specific discourses circulating among various actors about bioenergy and the public perceptions of bioenergy development. In particular, we want to emphasize the point that different types of wood-based bioenergy products are embedded in—and help to create—different imaginaries. The specific elements of these imaginaries are rearranged when considering, for example, whether wood-based bioenergy relies on experimental or established technologies; whether its end market is domestic or international; or whether its net carbon output is positive, negative, or neutral. These elements are directly tied to narratives about risk aversion, national security, energy independence, and climate change belief (or denial). Attention to these differences among production technologies and end products is one way of deepening our understanding of the role that imaginaries may play in local places where they have spurred concrete bioenergy development, places in which bioenergy is simultaneously literally produced and socially enacted.

PUBLIC OPINION OF BIOENERGY

There is growing recognition among various stakeholders (e.g., industry, researchers, policy-makers) that public opinion about bioenergy matters. Public opposition can delay or derail specific bioenergy projects, and weak public demand for bioenergy products can limit the development of policies friendly to bioenergy development. Skepticism about bioenergy's market potential can discourage landowners from changing their growing strategies for trees or potential new bioenergy crops. However, positive public statements about bioenergy from trusted sources may increase customer demand and foster technological innovation, financial investment, and political support for bioenergy. People's perceptions of wood energy development also vary greatly based on personal experience with, and knowledge of, the extant forestry industry and emerging bioenergy technologies.

Social science literature on wood-based bioenergy development in the southeastern United States has been limited. Several studies found low levels of understanding and misconceptions about bioenergy, suggesting the need for local outreach (Monroe and Oxarart 2011; Radics, Dasmohapatra, and Kelley 2016). Other studies examined perspectives of forestry professionals and forest owners, highlighting the importance of fitting bioenergy into existing wood production systems (Aguilar and Garrett 2009; Aguilar, Daniel, and Narine 2013). Bailey, Dyer and Teeter (2011) focused on the rural development potential of wood-based bioenergy development and the policies needed to ensure local benefits. If we view bioenergy promotion as a policy- and media-driven cultural phenomenon to meet a variety of energy, environmental, and rural development objectives, an important concern becomes the impact of national and regional energy and bioenergy imaginaries on local communities and landowners.

Bain and Selfa (2013) note that around 2008, bioenergy in the form of ethanol production from agricultural crops in the U.S. Midwest began to be questioned in terms of its ability to mitigate greenhouse gases; it began to be seen as having negative effects on food prices and food security. Burnham et al. (2017) described two wood-based bioenergy models, a regional approach and a community approach, that have emerged in the northeastern United States, with different subniche advocates contesting and shaping bioenergy imaginaries around issues of scale of development and land use. Kulcsar, Selfa, and Bain (2016) discussed the role of powerful interests in communities in shaping perceptions of bioenergy development as economically positive in spite of limited job creation, while at the same time downplaying negative environmental impacts. Eaton, Gasteyer, and Busch (2014) analyzed how actors in local communities in northern Michigan differentially framed national bioenergy imaginaries in support for or opposition to bioenergy development. They focused on local interpretations of the national bioenergy imaginary, specifically how a "wood for energy" scenario was differentially framed by actors in northern Michigan communities to either present it as an unproblematic, obvious choice or to emphasize risks, uncertainties, and complexities. Their analysis took an important step in asking how national sociotechnical imaginaries are interpreted and acted on in places where specific technological projects are unfolding (Eaton, Gasteyer, and Busch 2014).

Research on media framing of bioenergy is key to understanding perceptions of bioenergy, as media advance new discourses about bioenergy while also attempting to align their frames with the general public and thereby link bioenergy to larger cultural themes (Wright and Reid 2011). Wright and Reid (2011) claim that perhaps the single greatest influence on members of the general public is the media in its myriad forms: online stories, print newspapers and magazines, radio shows, documentaries, and social media. Popular media stories about biofuels often focus on the adverse effects of biofuels on the global food supply and use of arable lands for energy crops, as well as on high-profile scandals and failures in renewable energy such as Solyndra and Range Fuels (Hitchner and Schelhas 2012; Hitchner, Schelhas, and Brosius 2014). Wright and Reid (2011) and Sengers et al. (2010) note that shifts in the media framing of biofuels, from a positive focus on opportunities for economic and technological advancements to a more negative focus on environmental and social threats, have significantly contributed to public resistance to bioenergy. Dyer, Singh, and Bailey (2013) examined national, regional, and Alabama newspaper coverage and found that local coverage, in particular, is generally favorable when talking about bioenergy developments with potentially positive local economic effects. While conducting our own research, we have found that media coverage of bioenergy development in areas where new facilities will be located is generally positive, although failures are frequently cited in local media as well. Several media accounts with wide and diverse audiences have focused on the failures of bioenergy start-ups, including a December 2015 article in *Fortune Magazine* detailing the promise, then failure, of the KiOR cellulosic biocrude facility in Mississippi and the ongoing litigation following "the complex saga" of "how KiOR crashed so disastrously" (Fehrenbacher 2015). Although media frames may often be chosen to influence public and local opinion about bioenergy, we know very little about how they, along with the larger bioenergy imaginary, influence local people.

"Influencer culture" has permeated the energy realm as well, and certain individuals hold the power to influence millions of people. Celebrities including Willie Nelson, Bonnie Raitt, Neil Young, Morgan Freeman, and Claudia Schiffer have highlighted the ecological and social benefits of bioenergy, particularly of biodiesel and ethanol. Chuck Leavell, a keyboardist for the Rolling Stones, lives in the southeastern United States on a forest plantation and is an outspoken proponent of wood-based bioenergy. Onalytica (Fields 2018), a company that connects products to influencers on social media and other popular platforms, released an open-access truncated list of the top fifty influencers in energy in 2018; among the top ten were well-known politicians, activists, and entrepreneurs such as Andrew Cuomo (former New York state governor), Bill McKibben (350.org), Michael Bloomberg (former New York City mayor), and Elon Musk (Tesla). Actors Mark Ruffalo and Leonardo DiCaprio made the top twenty. As noted, politicians also have made influential statements about bioenergy and its place in the energy portfolio in the United States, though the issues related to bioenergy development cross party lines and can create strange bedfellows at times. For example, NASCAR (National Association for Stock Car Auto Racing) facilitated a collaboration between American Ethanol and the Richard Childress Racing Team to educate race fans, conventionally a politically conservative group, about the positive aspects of corn-based ethanol, drawing largely on values and discourses related to the importance of family farms. At one bioenergy conference we attended in Georgia in 2011, NASCAR driver Kenny Wallace, a spokesperson for American Ethanol, called himself "the voice of ethanol, the voice of American farmers"; he said that far from being "green voodoo," ethanol is a boon to small farmers and rural communities and completely consistent with a love for race cars. Another spokesperson for the NASCAR American Ethanol Performance Team, Kyle Mohan, has said that "Ethanol is a win-win. It delivers the power I need on the track, the reliability I need off the track, all while promoting a healthier environment and cleaner air" (Growth Energy 2021). By targeting the perceived support of NASCAR fans for both family farms and clean air, American Ethanol aims to gain their embrace of biofuels. Similarly, other public awareness campaigns on bioenergy have targeted specific groups of people, usually with the aim of promoting support for bioenergy development.

Environmental NGOs have expressed a range of positions, from staunchly opposing wood-based energy or expressing deep concerns about it (Dogwood Alliance, Natural Resources Defense Council, Sierra Club) to more measured views (National Wildlife Foundation) that see bioenergy as an energy option if appropriate measures are taken to ensure sustainable forest management. Meanwhile, several public health organizations, including the American Lung Association, American Academy of Family Physicians, and Physicians for Social Responsibility, have all publicly commented on the negative effects of wood-burning energy production facilities on human health. Other sources of information on wood energy within communities include local foresters and extension agents, industry representatives, friends, neighbors, landowners' associations, civic clubs, city and county leaders, and community interest groups.

All these individuals, organizations, and campaigns referenced above, in addition to many others, regardless of their credibility, reflect and sometimes shape ways of talking about and contesting bioenergy. Understanding local interpretations of bioenergy development, the interests and values that underlie these, and the ways that these lead to supportive or oppositional actions is necessary if bioenergy development is to be broad-based and collaborative, and to provide local benefits. Our goal in this book is to contribute to this conversation by analyzing the various themes that emerged in our ethnographic research, showing how people's sometimes incommensurable perceptions of bioenergy are influenced by their own experiences and by their interpretation of the imaginaries and powerful discourses they receive through the media and cultural influencers.

INFLUENCE OF DISCOURSES, NARRATIVES, AND METAPHORS ON PERCEPTIONS OF BIOENERGY

Curran (2012, 237) notes that "the force of storylines and narratives lies in their capacity to both simplify and organise discourses," and we believe that bioenergy metaphors similarly organize and simplify myriad discourses such as rural development, energy independence, national security, ecological sustainability, forest health, carbon sequestration, and government investment in technological innovation. These are often nested within larger discourses such as sustainable development or climate change and embedded in imaginaries, or holistic visions of current and potential realities (Fairclough 2010; Jasanoff and Kim 2009). As Jones (2014, 644–645) notes: "the historical record is littered with examples where narratives have been strategically generated by governmental actors to shape beliefs or restricted and contained to prevent them from doing exactly that." We contend that people are much more likely to remember words and phrases that evoke specific images or passionate emotions such as fear, anger, pride, or betrayal than they are to remember abstract concepts or technical data, especially those articulated in academic or technical jargon, and that in the case of bioenergy development, certain metaphors are circulated among various actors in ways that elicit emotional reactions. Recognizing the difficulty of proving that these metaphors directly influence people, we draw from two approaches to analyzing influence: conventional discourse analysis (Strauss 2012) and the notion of "moments of influence" (Witter et al. 2015).

Recent work on imaginaries has shifted away from viewing them as uniform and homogenous to paying more attention to processes of contestation and the ways that differing values and interests may produce alternative and incommensurable imaginaries and counternarratives (Delina and Janetos 2018; Delina 2018; Tozer and Klenk 2018). Similarly, Strauss (2006) cautioned against homogenizing and reifying imaginaries and stressed the importance of locating them in concrete actors, social contexts, and material conditions. Smith and Tidwell (2016) noted that Strauss (2006) and Jasanoff and Kim (2009) drew on different underlying intellectual traditions in their conceptualization of imaginaries. But, concurring with Strauss (2006), they advocated for greater study of the ways that sociotechnical imaginaries circulate more widely in society as they are "criticized, taken up, and reframed by ordinary people" (Smith and Tidwell 2016, 331). Strauss (2006) suggested that viewing imaginaries as structuring new ways of thinking is closely related to anthropologists' conception of cultural models as shared, implicit schemas of interpretation (Strauss 2006; Quinn and Holland 1987).

Cultural models are studied through text analysis, often of transcripts of talks with research subjects, sometimes through analysis of metaphors, and generally by identification of shared elements of individuals' mental models (Quinn 2005; Schelhas and Pfeffer 2008; Strauss and Quinn 1997). Notably, cultural models can be analyzed for diversity (some are coherent and shared, for example, while others show disagreement and conflict), as well as for motivating force (or lack thereof) (Strauss 2005; Strauss and Quinn 1997; Strauss 1992), which makes them well suited for looking for traces of national imaginaries in local discourse. Strauss (2012) coined the term "conventional discourse" to describe a type of cultural model—comments from multiple sources that convey the same assumptions using similar rhetoric to present a simplified view on a topic. Conventional discourses are compact mental schemas that are easy to repeat, and they represent accepted ways of thinking and talking within a particular opinion community (Strauss 2012); these qualities make them relevant to bioenergy imaginaries.

People often repeat common expressions that are linked to shared ideas in particular opinion groups, but these discourses are fluid; people unconsciously mix and match expressions from the variety of different opinion groups with whom they interact. Using conventional discourse analysis, Strauss (2012) explored the nuances of public opinion by showing how people mixed diverse ideological positions in complex ways when they talked about issues, drawing from different and sometimes competing public discourses or imaginaries that were or were not related to the topic at hand. Shared words and phrases that circulate within opinion communities constitute conventional discourses, and a speaker's use of them evokes a set of assumptions about what they think and value. Everyone belongs to multiple opinion communities, and people often express multiple and contradictory viewpoints about the same subject when drawing ideas and even specific rhetoric from different opinion communities.

The above discussion leads to the question of whether and how various aspects of dominant sociotechnical imaginaries are reconfigured and reimagined by individuals as they move into everyday talk, calling our attention to the experiences and agency of individuals in (re)interpreting and expressing concepts found in imaginaries. This can involve reshaping imaginaries to better reflect local interests (Smith and Tidwell 2016), the development of subniche imaginaries to promote different stakeholder visions (Burnham et al. 2017), or social movements that contest dominant imaginaries and promote new ones in broad public spheres (Kim 2015). But shifting to a people-centered approach based on cultural models and conventional discourses opens up an even wider range of possibilities. For example, Pfeffer, Schelhas and colleagues (Schelhas and Pfeffer 2008; Pfeffer, Schelhas, and Day 2001) used cultural models to analyze the different ways that local people seek to advance or defend their own livelihood interests in the face of powerful international conservation. The results were complex, including contestation, appropriation, and reinterpretation of discourses, as well as showing significant gaps between local talk and behavior (Schelhas and Pfeffer 2008; Pfeffer, Schelhas, and Day 2001).

Due to power differentials, everyday talk may have limited ability to shape sociotechnical imaginaries themselves but may still indicate everyday resistance that inhibits the necessary collaborative and multi-scale efforts that bioenergy development requires. As state-originated sociotechnical imaginaries lead to material developments, the ultimate fate of both the imaginaries and material developments will depend on and reflect the ways that people talk and act in local places.

While Strauss (2012) focused on how opinion communities influence public perceptions of complex social issues such as immigration and welfare, Witter et al. (2015) examined the question of how to determine the level of influence that certain actors have on large-scale political processes. In observing the potential influence that indigenous actors had on negotiations at the 10th Conference of the Parties of the Convention on Biological Diversity (CBD COP10) in Nagoya, Japan, in October 2010, they draw a distinction between influence and outcomes, noting that influence can be both "spontaneous and elusive, but nonetheless strategic and tactical" (Witter et al. 2015, 3). In using the word "moments," they emphasize that much of the influence that actors have on policy is "situational and incremental, and thus might be invisible to those not observing the negotiations in real time" (Witter et al. 2015, 2). In other words, even if the objectives of certain actors are not met in any one set of global policy negotiations, their actions over time can have strong impacts in other arenas later. Ethnographic attention to these "moments of influence," which may not have immediate consequences but that contribute to later changes, is one way to analyze long-term influence by observing behavior and interactions among various actors.

In the context of work by Strauss (2012) on conventional discourses and by Witter et al. (2015) on "moments of influence," we explore the impact that specific metaphors can have on perceptions of bioenergy. Quinn (2005) argues that metaphor analysis is a particularly useful method of inquiry to elucidate shared cultural understandings of complex social institutions such as marriage because people frequently use metaphors in order to clarify the points they are trying to make, and others within the same cultural context readily understand the metaphors and the meanings behind them. We show how an analysis of metaphors complements a multisited ethnographic approach to understanding various actors' perceptions of wood-based bioenergy, as metaphors circulating among and between different actors crystalize key messages about the potential risks and benefits of bioenergy development. We contend that the use of these metaphors by different actors in different settings (written and online media, in speeches and presentations at bioenergy conferences, and in casual use among members of the general public), act as "moments of influence" that reinforce cultural values such as national sovereignty, environmental protection, and economic success.

Combining analysis of conventional discourses and metaphors reveals both cultural meanings and memorable ways of symbolically translating and transmitting these meanings among actors operating at different scales. We suggest that metaphors are used in two ways in bioenergy discourse:

(1) consciously and strategically, with the explicit intent to influence; and (2) unconsciously, as a means of passing along shared ideas. As stated by Lakoff and Johnson (2003[1980], 139) metaphors can "give new meaning to our pasts, to our daily activity, and to what we know and believe." They note that metaphors, and interpretations of them, are specific to the cultures, life experiences, and physical environments of the people that use them. Building on these ideas, Quinn (2005, 49) explains that people use metaphors as tools of clarification when speaking to others within their own cultural group (who will readily understand them), and that metaphors serve as "cultural exemplars," or "particularly salient intersubjectively shared examples of what they stand for." Our examination of creative metaphors that reference specific conventional discourses can help us to identify some of the translational dynamics and incommensurabilities in the ways that these phrases and ideas travel within and between different opinion communities and act as moments of influence on public perceptions of bioenergy.

THEMES, OR ELEMENTS OF IMAGINARIES

Discussions about bioenergy take place on multiple levels simultaneously. While a coherent and optimistic bioenergy imaginary in which bioenergy solves multiple problems is articulated, debated, and contested at the national level, different actors at the local level are hearing and sharing various elements of these national discussions in the context of their own experiences. It is not clear to what extent local experiences move upward to influence national imaginaries, but we argue that local experience may shift the meaning of national imaginaries and change the way they influence behavior. In this book, we present results from our ethnographic research that show how elements of imaginaries were promoted at those sites and then how they were expressed in a variety of ways through conventional discourses. These main elements, which will be explored in depth in the following chapters, include: (1) energy independence and security, (2) rural development, (3) government promotion and investment, (4) climate change, (5) forests and forest products, and (6) renewability and sustainability. These form the building blocks of bioenergy imaginaries, and we show later how these elements are interpreted differently by different actors. In the following section, we show how a few recurring metaphors encapsulate many of these themes and are passed on among various actors, both in an "everyday talk" sense in which people simply repeat what they hear and in a strategic sense where speakers are intentionally invoking emotion for the purpose of persuasion.

METAPHORS: SHORTCUTS TO IMAGINARIES

Building on these main themes that we encountered in our fieldwork, we decided to analyze them with a focus on three prominent and evocative metaphors. Again, we emphasize that while that some actors use these metaphors strategically to influence public opinion on wood-based bioenergy, others simply pass them along like other bits of speech that they remember because they personally resonate. They include:

1. "snake oil," referring to the shady practices of some people promoting and presumably profiting from bioenergy development;
2. "silver buckshot," referring to the idea that bioenergy is not a singular object or process, and thus involves many different technologies; and
3. "people who hate us/want to harm us," referring to the idea that companies and government agencies in the United States buy petroleum from countries that have initiated terrorist attacks against U.S. citizens.

We heard many other metaphors and phrases repeated throughout our fieldwork. A few examples include: "low-hanging fruit," referring to easily obtainable renewable energy goals; "Holy Grail," in reference to drop-in liquid fuels that do not require infrastructure changes for their widespread use; and "killing the goose that laid the golden egg," referring to the fear that development of woody biomass as a bioenergy feedstock will harm well-established forest products industries such as lumber and pulp and paper. However, we have chosen to focus on these three because they encapsulate many ideas that drive the perceptions of people in the southeastern United States about bioenergy and are easily linked to some of the most common conventional discourses that we encountered (for more detail, see Hitchner, Schelhas, and Brosius 2016 and Schelhas, Hitchner, and Brosius 2018). They were also the metaphors that we heard most often during fieldwork, and they symbolize the conflicting imaginaries perpetuated by different bioenergy stakeholders.

Snake Oil

The phrase "snake oil" refers to traveling salesmen in the 1800s selling elixirs that they claimed could cure any ailment. The phrase is commonly used today to refer to fake products, or products that make false claims; people that sell or promote such products are called "snake oil salesmen." People who realize that they have purchased snake oil feel embarrassment and shame for allowing themselves to be fooled, as well as anger at the salesman who betrayed their trust. This metaphor has been applied to a multiplicity of topics in many disparate academic disciplines, and it is similarly used in many contexts in popular media.

There have been many applications of this metaphor to bioenergy as well. While there have been historical shifts in the discourses about bioenergy, from panacea to highly problematic, many proponents of bioenergy continue to promote bioenergy as a "win-win" (or sometimes as a "win-win-win") (Scott et al. 2014). Bioenergy development, for example, will contribute to national security, energy independence, rural development, restoration of "idle" or marginal landscapes, reduction of greenhouse gas emissions, and will spur technological innovation and create jobs in the United States that cannot be outsourced. Opponents or skeptics of bioenergy rebut each of these points, referring to the same conventional discourses. When bioenergy is touted as a win-win, it is comparable to the "cure-all elixirs" sold by traveling salesmen. When there are highly publicized bioenergy failures, especially involving government/taxpayer money, some people feel that the companies who promised—and then failed—to deliver a bioenergy product are "snake oil salesmen." This implies that the companies (and often the government as well) were fundamentally dishonest, knowing that the product was fraudulent. Like the sellers of snake oil liniments that planned to be gone by morning and selling in another town the next day, some bioenergy companies are accused of being "fly by night" operations that never intended to create any product at all; they only planned to collect money from government subsidies, incentive programs, and investors, in addition to the "goodies" offered by hosting states and counties. We contend that this image and the emotions that it evokes, when applied to bioenergy, are linked to several conventional discourses such as government corruption, industry greed, waste of taxpayer money, and unfair policies that pick "winners and losers."

In general, this metaphor calls attention to public perceptions of not only industries as potential snake oil salesmen, but also to the fairness of bioenergy policy and the role of the government in supporting bioenergy producers. The question of who pays for, and who benefits from, bioenergy development is at the forefront of many people's concerns; several studies have revealed how members of the public have expressed profound misgivings about how bioenergy policy shapes the development of bioenergy technologies and facilities. For example, Delshad et al. (2010, 3420) found that study participants generally approved of government initiatives supporting bioenergy development, including subsidies and other incentives for producers who were at a disadvantage in the market and were taking financial risks that could benefit society, even while opposing government involvement, "which they saw as an unfair and unacceptable intrusion into the free market." Taxpayer-funded subsidies given to developing industries to meet government mandates for renewable energy are sometimes perceived as closed-door deals between politicians, lobbyists, and CEOs. Bryce (2008, 11) says that "ethanol is one of the biggest frauds ever perpetuated on U.S. taxpayers," and our research results show that many people in the southeastern United States agree with him.

Around the time that we were conducting fieldwork references to bioenergy as "snake oil" abounded in written documents available to the public. For example, in 2013, Captain T.A. "Ike" Keifer, a naval aviator and teacher of strategy at the U.S. Air Force Air War College, published a report called "Twenty-first century snake oil: Why the United States should reject biofuels as part of a rational national security energy strategy." From 2006 to 2013, several other online sources published articles and blog posts with titles such as: (1) "Is Solazyme selling investors snake oil?" (Nanalyze 2013); (2) "Ethanol: The latest incarnation of snake oil" (Scheurich 2010); and (3) "Cellulosic ethanol—snake oil for the new millennium?" (Miller, 2006). The *Denver Post* ran an article in 2007 called "Ethanol is politicians' snake oil" (Harsanyi 2007). A peer-reviewed article in the *Oregon Law Review* was titled "Biofuels: Snake oil for the twenty-first century" (Reitze 2009), while the nongovernmental organization FERN (which is focused on social and environmental justice issues regarding forests and forest peoples in Europe) created a publication in 2009 called "Snake oil or climate cure: The effect of public funding on European bioenergy" (Fenton 2009).

We also encountered specific references to "snake oil" during our ethnographic fieldwork, either using those words directly or invoking the metaphor indirectly using words that carry essentially the same meaning and evoke similar emotions such as "scam," "bill of goods," "hoodwinked," "shady deals," or, promises to "weave straw into gold." A forester near Waycross, Georgia, said, "People see it [bioenergy] as snake oil; they're wary of it. You have to show them it'll work." In response to a question on energy grass, an employee of the Natural Resources Conservation Service (NRCS, which operates under the U.S. Department of Agriculture) working near Soperton, Georgia, stated that "people see *Miscanthus* as snake oil." Another government employee, with the Resource Conservation and Development Program of the NRCS near Soperton said of Range Fuels, "It never did happen. They hoodwinked us all on that one." A county leader near Soperton said, "I have seen reports after the failure of Range Fuels. I heard that it was a scam," while a community member in Soperton stated that after the failure of Range Fuels, "the county got left holding the bag." A forester near Soperton said, "Range Fuels sold us a bill of goods. They had all the technology, just had to use it. They didn't have as much as they said they had." A farmer in the same area said, "The government doesn't need to be in that business [bioenergy]. There's a lot of shady deals, people stuffing their own back pockets . . . If it was going to work, private investment should take care of it." In Columbus,

Mississippi, where the KiOR plant failed, one community leader described KiOR this way:

> We call it a "house in a box." One of those bullshit projects that is never intended to do anything but make someone rich. You know when there's a storm and people are homeless and someone promises to build them a house in a box, and it's nothing, you know, nothing. It's too good to be true. So shit like this [KiOR], we call it a house in a box. Too good to be true. They say they're weaving straw into gold and you know they can't.

A community member there stated that "this whole thing was a $75 million experiment funded by the state of Mississippi." On a higher level, a commenter at a bioenergy webinar facilitated by the U.S. Department of Energy in 2011 stated that "If you want public support, it will only be maintained if there are real benefits being delivered. Not just a public perception thing—there must be actual benefits. You can only pull the wool over their eyes for so long." Our research has confirmed that this is very true. The metaphor "snake oil," and the ways that we have observed various actors in the southeastern United States apply it to bioenergy, encapsulates several traditional American values and conventional discourses such as policy fairness, honesty and trust, and governmental support for both entrepreneurs and "the little guy."

Silver Buckshot

The phrase "silver buckshot" conjures the image of a werewolf, which, according to popular folklore, can only be killed by a bullet made of silver. There are several possible origins for the legend of the werewolf. The first is rooted in Greek mythology, where the Arcadian king Lycaon was turned into a wolf by the god Zeus after Lycaon tried to serve him a meal of human flesh. A second popular origin story dates to the mid-1760s, when the Beast of Gévaudan, a giant wolf-like creature, terrorized the rural French countryside, feasting on shepherds, with a special appetite for the fat of children (Smith 2011, 20). Jean Chastel, the hunter credited with killing this beast, claimed to have used a bullet made from a molten silver chalice that had held the blood of Christ and had been blessed by a priest (Steiger 2012, 252). The producer of the 1941 film *The Wolf Man* claimed to add many details to the werewolf myth, including the fact that they can only be killed by silver bullets (Steiger 2012, 252). In any case, as with many myths and legends, stories were embellished and woven together over time by different storytellers, and werewolves remain a prominent figure in popular literature, cinema, and folklore around the world.

The idea that there is no "silver bullet," or singular solution, to solve a complex problem has been applied to many issues; the term is probably most commonly applied in academic literature regarding climate change. Gun metaphors are pervasive in American culture in general (Jones 2014), and Gerhart (2010) provides numerous examples of politicians using the phrase "silver bullet" in reference to a number of societal issues, noting that during his presidential campaign, Barack Obama used the phrase over sixty times. The metaphor "silver buckshot," an extension of the "silver bullet" metaphor, is commonly used to refer to a number of complementary mechanisms needed to address complex problems such as climate change. The phrase is also used by people in the bioenergy sector to express the idea that meeting global or national energy needs will require energy from many sources, including fossil fuels and from many types of renewable energy sources, including woody biomass, biogas, wind, solar, hydroelectric, geothermal, and wave/tidal energy (Mittlefehldt 2016). Woody biomass is one piece of silver buckshot that will help to kill the metaphorical werewolf (which could be dependence on fossil fuels or an inadequate supply of energy), but it will not be the silver bullet that can take it down alone.

In the bioenergy world, the phrase "silver buckshot" also refers to the idea that there is no one particular technology, conversion process, or feedstock that will be the salvation of using biomass as an energy source. Different technologies, processes, and feedstocks are likely to work better (or worse) in different places and at different scales, and failure of some emerging technologies is often a necessary part of eventual success. People working within the bioenergy sector also refer to their processes as being "technology or feedstock agnostic," meaning that the technology can use different feedstocks, or that the policies do not choose "winners and losers," but instead offer a "level playing field" to different actors. Again, the idea is that there is no one solution to the problem of how to effectively, efficiently, and equitably supply the world (or the nation) with enough energy, particularly renewable energy. Here we apply the metaphor "silver buckshot" to the variety of energy sources (real and potential), technological pathways to produce bioenergy, and levels of bioenergy regulation.

We found several sources using the phrase "silver buckshot" in written publications starting around 2010 in reference to an approach to an energy portfolio that includes bioenergy, including that derived exclusively from woody biomass. In 2010, PBS (Public Broadcasting Service), which aims to reach a wide public audience, published an opinion piece called, "Energy's silver buckshot" (Fri 2010), referring to the need for a diversified energy strategy. In 2012, an article in *Biomass Magazine*, a publication geared to bioenergy industry insiders, focused on wood pellets as one piece of the bioenergy puzzle; its title was "Biomass' own silver buckshot," referring to wood pellets as one piece of the bioenergy puzzle (Portz 2012). Another trade magazine, focused on transport, published an article in 2013 called "Carriers use 'silver buckshot' to shoot down fuel costs" (Long 2013). The popular magazine *Wired*, which focuses on the intersection of science and technology in daily life, ran an article in 2012 called "A 'silver buckshot' will mend our fuelish ways" (George 2012). In 2018, a peer-reviewed article in the academic journal *Sustainability* was titled "Silver buckshot or bullet: Is a future 'energy mix' necessary?" (Brook et al. 2018). The phrase has clearly gained and maintained popularity in academic, trade, and popular sources (for more examples in relation to bioenergy, see Hitchner et al. 2016).

In our own fieldwork, we found that the phrase "silver buckshot" was used by several people at bioenergy conferences; these are the people most likely to have come across the phrase in the trade publications and online sources referenced above. At one bioenergy conference in 2011 in Galax, Virginia, a participant stated, "Biomass is not a silver bullet; it's more like silver buckshot." A forester at the same conference, clearly a supporter of wood-based bioenergy, referenced the first speaker by saying:

I'm a forester, and I see that biomass is often not utilized, and that landowners are not managing their hardwoods and pines. That's what fueled this approach . . . In 1997/98, fuel prices were high, and I knew those high prices would come back. This solution is that one piece of silver buckshot we talked about yesterday.

At another bioenergy conference in 2011 in Tifton, Georgia, a speaker stated, "There is no silver bullet, we must develop a silver buckshot vision." The same idea was phrased slightly differently at another bioenergy conference, this one an online webinar hosted by the U.S. Department of Energy: "Finally, we need an all-of-the-above energy policy. We can't rely solely on natural gas; that thinking is fundamentally flawed. We can still promote renewable energy sources even if it's more expensive than natural gas." This phrasing references President Barack Obama's "All-of-the-Above" Energy strategy dating from 2016 (which was later modified by the Trump administration to focus on fossil fuels with minimal attention to any kind of renewables). An interviewee in Tunica, Mississippi, repeated the phrase as well: "We've got markets, got veterans who need jobs. Regional bioproducts could do that. It's not the full answer, but one part. It's an 'all-of-the-above' situation." A speaker at a bioenergy conference in Tunica, Mississippi, also mentioned bioenergy from woody biomass at the then-proposed KiOR facility in this way, "Trees don't grow fast enough to meet all energy demand. We need a portfolio. Does KiOR help? Yes." KiOR, then, could become one piece of buckshot; it ultimately was not, but the potential was there.

People Who Hate Us

For many Americans, the expression "people who hate us," or variations of it, evokes the image of terrorists who hate the United States and want to harm or kill Americans. With this image, usually of Muslims of Middle Eastern descent flying planes into the World Trade Center and the Pentagon on September 11, 2001, comes the idea that the dependence of the United States on foreign sources of petroleum makes the country vulnerable. There is much talk of the "U.S. addiction to foreign oil" and the desire to be free from this dependence and therefore from people who want to harm the country and its people. This "culture of fear" (and accompanying xenophobia) is also perpetuated by mainstream media, particularly politically conservative media outlets such as Fox News and right-wing radio shows.

A number of written publications have titles which include the phrase "people who hate us" or some variation. In 2008, a contributor in a thread on an electric car website wrote, "Stop giving money to people who hate us: Anti-oil mud-slinging festival" (Anonymous 2008), while a 2008 letter to the editor in the *Chicago Daily Herald* was titled "Buying oil from people who hate us" (Wagner 2008). The *Center for American Progress*, a non-partisan progressive policy institute, published on its website an article entitled "Oil dependence is a dangerous habit" (Lefton and Weiss 2010), referencing the fact that most of the oil consumed in the United States comes from politically unstable countries; they further note that climate change caused by burning fossil fuels also threatens national and global security.

During our research, we encountered numerous references to the idea that buying fossil fuels can increase reliance on countries hostile to the United States. At a bioenergy conference in Tunica, Mississippi, a natural gas representative talked about how new technologies make locally harvested natural gas a clear alternative to purchasing fossil fuels from other places; he said, "Natural gas is American made, and cheaper. It's come into our laps because of new technologies . . . . What we need is national security . . . Before fracking, we were about to import all our natural gas. Now we can buy less oil from people who don't care for us." Another speaker at the same conference who clearly favored wood-based bioenergy over fossil fuels of any kind, said that by using locally grown woody biomass to produce energy, "we're using American resources to make American fuel instead of buying from people who want to kill Americans."

At another bioenergy conference in Tupelo, Mississippi, one speaker stated, "Why biofuel? Dependence on foreign oil . . . 40% of U.S. oil consumption comes from imports. There's much political unrest in supplying countries, especially the Persian Gulf . . . Less energy security equals more vulnerability." Another speaker there said, "Sustainability means creating an environment for security and having the resources to do it . . . Terrorist, failed and rogue states—we're more connected to them now; they'll always have global impact."

This idea was also referenced a number of times in an online webinar organized by the Department of Energy. Tom Vilsack, secretary of agriculture, said, "We've got to move away from our addiction to foreign oil . . . The Gulf is not a reliable source; those countries are in turmoil . . . We buy oil from people who would do us harm . . . We have to tap our own resources." Jackalyne Phannenstiel, secretary of the Navy, said,

> Why is this energy mission critical to the Navy? The first reason is energy security; we need to stop buying fuel from unstable and hostile places. There's a risk for our military to transport gasoline; there are dangerous fuel convoys in Afghanistan and other places.

Finally, Senator Amy Klobuchar directly referenced the terrorism that resulted in the attacks of September 11, 2001, by stating,

> Ignoring energy issues is undercutting U.S. foreign policy . . . Where could we have moved on this? The first moment was after 9/11, when the whole nation was horrified and united. It was the right decision to invade Afghanistan, but we could have reduced our reliance on those countries that produce terrorism.

This phrasing, and the underlying emotions it evokes, is commonly referenced when people discuss U.S. imports of petroleum from other countries. We contend that this metaphor demonstrates the power that fear-based politics has in the United States, as well as evoking American values of national pride, independence, justice, resilience, and cultural sovereignty or national isolationism.

LINKING METAPHORS AND CONVENTIONAL DISCOURSES

As noted, people are more likely to remember words and phrases that evoke clear images and strong emotions than abstract concepts. People have been using these types of rhetorical devices for millennia as both mnemonic aids and as means to strategically influence people. It is logical that people are more likely to remember an image of a werewolf that is shot by a shotgun full of buckshot (instead of the usual single silver bullet from a rifle or pistol) than they are a lengthy, jargon-filled explanation of the multiplicity of manifestations of renewable energy alternatives (or complements) to fossil fuels and of bioenergy feedstocks and processing technologies. They are also more likely to be stirred by anger at being "hoodwinked" by corrupt companies or "swindled" by the government "at taxpayers' expense" by "snake oil salesmen" than to remember specific statistics about how much money is involved from what sources. And the images of people jumping from the collapsing Twin Towers or the burning hole in the Pentagon, the ultimate symbol of national security, are much more likely to resonate than a dispassionate discussion of the goal of attaining a more diversified national energy portfolio.

It is important to note that people just used these or similar phrases; they didn't graphically describe these images. But the images are evoked, even if they're not spelled out in detail; human imagination fills in the blanks. It is also important to recognize that some people are more creative and imaginative than others (and thus more likely to use and appreciate metaphors) and that many metaphors are highly culturally specific, so metaphors are not universally relevant. Metaphors are also drastic oversimplifications of very complex issues, so reliance on such images strips much of the nuance and critical analysis relevant to a comprehensive discussion of a complicated topic such as alternative energy. We contend that when actors use such metaphors intentionally, they do so strategically in order to not only condense meaning but also to promote a specific normative slant that reinforces a specific imaginary in which wood energy does or does not act as a solution to current problems. At the same time, they often act as translational vessels that people deploy in their efforts to speak across difference. However, the way that such images and emotions are evoked in particular geographic and cultural contexts is an important question for cultural, as well as cognitive and linguistic, anthropologists.

Here we propose a framework for connecting metaphors and colloquial expressions with images and emotions and then linking these to conventional discourses that are passed on through different opinion communities. The metaphor "snake oil" conjures a mental image of a traveling salesman; associated rhetoric includes words like "scammed," "hoodwinked," "taken to the cleaners," and "sold a bill of goods." These phrases all evoke negative emotions such as anger and betrayal. They can be linked to conventional discourses such as government corruption, industry greed, and uneven rural development, and they circulate within opinion communities such as media, NGOs, community leaders and members, and landowners. "Silver buckshot" as a metaphor brings to mind the image of a werewolf who can be destroyed not just by one silver bullet, but by a combination of smaller weapons. Associated phrasing includes "All-of-the-above strategies," "[no] one-size-fits-all solutions," and "feedstock/technology agnosticism." These ideas evoke more positive emotions such as hope, creativity, persistence, drive to succeed, and cooperation. These ideas fit into conventional discourses such as diversified portfolios, alternative solutions for alternative situations, and persistence and cooperation and American values. We witnessed these ideas being promulgated within opinion communities such as politicians and actors within the bioenergy and forest products industries. Finally, the metaphor "people who hate us" creates in the receiver's mind pictures of terrorists. These are alternatively described as "people who want to kill/harm us," "people who don't like us/care for us," "powder kegs," "clear and present dangers," and "rogue states." This metaphor conjures a number of emotions, both positive and negative; it can trigger feelings of fear, distrust, uncertainty, xenophobia, and racism, as well as national pride, self-reliance, and unity in the face of danger. Opinion communities such as politicians, political pundits (especially right-wing ones), military personnel, and actors within the bioenergy industry circulate this metaphor as a way to persuade people to support locally produced bioenergy. While there is cross-pollination here of ideas, metaphors, emotions, influencers, and discourses as they travel and are translated across circuits of influence, these three and other metaphors encapsulate many of the ideas circulating in the public sphere and within the social circles of various industries, federal and state agencies, and nongovernmental organizations.

The power of rhetorical devices to influence people has been understudied, and perhaps even underestimated, in research focused on how information, narratives, imaginaries, and metaphors related to energy in general, and bioenergy specifically, are shared and perpetuated among people. This chapter presents a conceptual model for organizing the major conventional discourses that appear in discussions of bioenergy in the southeastern United States and for linking them to widely circulated metaphors and images that evoke clear images and trigger strong emotions. This model organizes and clarifies data obtained during multisited ethnographic research with communities in the southeastern United States experiencing the effects of bioenergy development and participant observation with bioenergy industry networks. It also illuminates broader patterns in wording, translation, incommensurability, and contestation related to renewable energy technologies and the normative implications of the circulation of these particular metaphors on members of the general public.

Forest-dependent communities in the southeastern United States have experienced a long history of boom-and-bust cycles in wood products and of failed promises to bring new economic development and employment opportunities to rural communities. The metaphors of "people who hate us" and "silver buckshot" serve to rationalize and adjust expectation for the process of developing a new market by saying we need to do it and that we need to try many things to be successful. In applying this conceptual model linking conventional discourses and imaginative metaphors to our ethnographic research in this context, we can see how these images are applied to specific instances of failures of bioenergy projects (such as Range Fuels and KiOR). These failures have not only cost the states, counties, and towns in which these facilities are located millions of dollars in grants and loans. They have also shattered the confidence of local community members and landowners in bioenergy as a potential solution to rural poverty and have led rural people already wary of government intervention in industry to feel even more betrayed by government policies that seem to benefit a few at the expense of the public in general and local communities in particular. "Snake oil" crystalizes the opposition to bioenergy that has emerged out of these failures and a general distrust of government intervention in markets.

It is also important for policy-makers, politicians, and local community leaders to recognize that these failures, widely publicized in print, online, and social media, have also had reverberating effects on public opinion outside these forest-dependent communities, in addition to dire financial consequences for bioenergy investors and entrepreneurs. Many people with whom we spoke representing local governments and communities felt they were sold "snake oil" by traveling salesmen who did not have to suffer the consequences of these failures. However, bioenergy representatives and many local and national politicians are still optimistic about the potential role that wood-based bioenergy can play in a sociotechnical imaginary that brings together carbon-neutral, sustainably sourced energy and rural economic development. They talk about wood-based bioenergy in the southeastern United States as one piece of "silver buckshot" that can help slay the werewolf of dependence on "terrorist" nations full of people "who want to harm us." The fundamental disconnect between the negative view of bioenergy as "snake oil" and the positive view of it as a piece of "silver buckshot" that can be used as a powerful weapon against "people who hate us" reveals the divergence of imaginaries in which wood-based bioenergy is, or is not, a potential solution to current and future energy problems. It also reveals differences in personal experiences with bioenergy in local contexts and policy arenas; for example, people promoting bioenergy through industry development or policy creation often do not live in rural communities in which bioenergy facilities are located and thus do not feel the effects of these facilities' success or failure.

In the next chapter, we explore in more depth the way these various imaginaries collide in rural, forest-dependent communities in which wood-based bioenergy facilities are established. Here in these places, we see the collisions and entanglements of various narratives and imaginaries, as well as the reality of differential benefit distribution on different subpopulations, in a region where the forested landscape itself is deeply imbued with cultural meanings and sentimental attachments.

NOTE

1. The Renewable Fuel Standard (RFS) was enacted in the United States in 2005 and expanded in 2007 to ensure that fuels made from renewable sources would be blended into transportation fuels such as gasoline and diesel in order to reduce carbon emissions to address climate change, lower fuel prices, and reduce the country's dependence on foreign oil. The percentage of renewables mixed into gasoline had previously been capped at 10%, though in 2011, the blend was increased to 15% for cars built in 2001 or later; as expected, this angered the fossil fuel companies as it boosted the biofuel producers. Cars being produced now can handle higher concentrations of ethanol, and this trend is expected to increase.

Chapter 3

Bioenergy Landscapes

Impacts of Bioenergy Developments on Forest-dependent Communities in the U.S. South

THE FORESTED SOUTHEASTERN UNITED STATES AS A CULTURAL LANDSCAPE

The southeastern United States is a cultural landscape with long overlapping histories of human occupation and resource use. Forests have always been a primary biome in this region: the dominant ecosystem before large-scale clearing for agricultural plantations, the site of monocrop tree plantations on cleared natural forests or former agricultural land, and many thousands of acres of successional forests (and some remaining old growth) that are preserved by individuals, governmental agencies, nongovernmental organizations, Tribal nations, and land trusts. The southeastern United States contains 40% of the nation's timberland (Oswalt and Smith 2014), and accounts for over 60% of the timber harvested in the United States (Oswalt et al. 2014). Ninety percent of the timberland in the southeastern United States is privately owned, with family forest owners controlling about 63% of that acreage (Wicker 2002). Rural communities within these forested regions have long relied on income from forest products such as lumber, turpentine, pitch and naval stores, pulpwood, mulch, and pinestraw. There are numerous communities in the South that continue to rely on forest products as drivers of employment and revenue, and often their community identity centers on a long history of forestry. The decline of wood products industries in rural communities has created an economic and social vacuum and has fostered enthusiasm among some communities and forest landowners for filling the gap with new wood markets, including wood-based bioenergy. However, this enthusiasm is tempered with concerns about ecology, community cohesion, industry greed, and political whim.

We explored energy, in particular, wood-based bioenergy, imaginaries—and the various ways that people articulate ideas about them—in the previous chapter. Here we aim to situate these imaginaries in context, in real places on the ground. We present some of the numerous ways that different actors imagine forests: as commodified resources, as ecological havens with intrinsic value, as repositories of human legacy and memory, or even as sacred places imbued with divinity. We tie these different visions of what forests are to ideas about how various imaginaries configure the role of wood-based bioenergy in the future of the region's forests. As we will see, these visions of forests and the energy they can produce are simultaneously divergent and incommensurable, and overlapping, even within individuals who live here.

Race has always been an integral part of the forested landscapes of the U.S. South, whether as vast expanses stolen from Native American tribes, as hideouts for enslaved plantation runaways, or as current landholdings entangled with legal issues regarding heirs' property. Racial issues, including environmental justice concerns, continue to be an important consideration in the development of wood-based bioenergy facilities. We explore these issues in more depth in chapter 5, but they cannot be left out of this discussion about forests as cultural landscapes.

Building on the previous chapter, we present here how different people talk about issues surrounding the development of wood-based bioenergy; they do this in both casual and strategic ways, and the language they use may or may not lead to productive communication and equitable decision-making processes. We demonstrate how winners and losers are inevitable as a result of bioenergy policies, which actors often frame as a win or a loss for the forested landscapes on which these policies are enacted. Again, these incommensurabilities are neither uniform nor static, as technologies and policies are constantly in flux and as the complex social and economic dynamics vary widely both within and between communities.

IMAGINARIES IN REAL PLACES: THE FORESTED CONTEXT

Forests in the U.S. South can be regarded as complex socio-ecological systems in which forest characteristics have long reflected human activities. Climate and vegetation changed as glaciers retreated following the late Wisconsin glaciation, both up to and after 10,000 years BP (Carroll et al. 2002). Native Americans altered the southern environment, primarily with fire, for at least the past 12,000 years, creating the mosaic of open and uneven-aged forests, Native settlements, agricultural lands, and prairies encountered by early European settlers. Forests subsequently changed—with agricultural systems turning to canelands and forests—following the decline of Native populations due to diseases introduced with European colonization. The South was dominated by fire-created forests such as longleaf pine, even though without fire more shade tolerant species would have been present (Carroll et al. 2002; MacCleery 2011). Subsequent to European colonization, settlers used fire to clear land for homes and farms, particularly along the terraces of large river systems and on the coastal plain (Owen 2002). Forests were important for and shaped by resin extraction, then clear-cut logging, and ultimately reforestation and forest management (Owen 2002). Prior to the advent of fertilizer, "old" southern fields were often allowed to regrow into pine (Williams 1989), after the decline of large-scale row cropping of crops such as cotton and tobacco. Further changes in agriculture, including mechanization; conglomeration of farms (and subsequent loss of many family farms); the introduction of genetically modified crops; more intensive growing methods made possible by stronger chemical-based fertilizers, pesticides, and herbicides; and geographic shifts of large-scale production of crops such as cotton and tobacco overseas, all contributed to a decline in the acreage of land used for farming. This former farmland has often been converted to trees, either through natural generation or intentional planting, on land held by families. Increased demand for pulpwood and timber by the wood products industry has led to the creation of large-scale tree plantations by the timber industry or by private landowners under government assistance programs.

For centuries, the forests of the southeastern United States have provided economic and ecological benefits, and forests are an integral part of the cultural and social identity of many local communities in the region. Forest products industries, ranging from turpentine production to whole-log export, have contributed significantly to changes in local economies, communities, and landscapes. Over the past several decades, outmigration of young adults, mechanization in agriculture, conversion of agricultural lands to forests, and declining forest industries have played a role in the weakening of economies in many parts of the rural South. In the southeastern United States, the economies of many communities have been built on forest products industries, mainly utilizing native pine trees; the abundance of pines here is a driving force of bioenergy development in this region. To date, much of the woody biomass used in the bioenergy industry has been pulpwood made available by declines in the pulp and paper industry (Brandeis and Guo 2016; Poudyal et al. 2017). But bioenergy research has also focused on new species, including hardwoods and eucalyptus, and silvicultural changes, such as shorter rotations, changing planting densities, and new mixes of genetic material (Hinchee et al. 2009; Nettles et al. 2015; Lu and El Hanandeh 2017). There is little doubt that bioenergy has the potential to result in new patterns of forest cover.

Within this dynamic context of ongoing change and regrowth, the emergence of wood-based bioenergy facilities in forest-dependent communities has led to both promises of new wood markets and disappointments when not all of these promises have materialized, perpetuating the cycles of enthusiasm and disillusionment described in previous chapters. In the early 2000s, as noted, a bioenergy imaginary emerged that solidified the potential of wood-based bio-energy as a means to simultaneously address climate change, achieve energy independence and security, and promote rural development in economically depressed forest-dependent communities in the southeastern United States. Existing commercial-scale wood pellet production for the European renewable energy market and a nascent liquid biofuels industry using wood and grasses as feedstocks have been enthusiastically embraced in many sectors of this region.

While this dominant bioenergy imaginary promulgated by the government and powerful interest groups has spurred bioenergy development, it has not been universally celebrated. Incommensurable counternarratives have circulated in this region, often originating from environmental NGOs and advocates who claim that bioenergy markets using woody biomass as a feedstock will lead to large-scale deforestation, landscape fragmentation, and increased air pollution. Concerns about the effects of wood-based bioenergy developments on established wood markets, local economies, and the health of both people and ecosystems have, in some cases, fueled skepticism about or even direct opposition to bioenergy development. Community members, citizen groups, nongovernmental organizations, and forest industry representatives sometimes oppose construction of new bioenergy facilities in rural, forested areas. While different apparently incommensurable imaginaries can occur simultaneously without much apparent interaction, they usually share common elements seen from a variety of perspectives.

Local community members who have experienced the economic, ecological, and social effects of bioenergy development firsthand are in a unique position to comment on how these imaginaries play out on the ground in areas where bioenergy is sourced and produced. As noted in the last chapter, local communities in which bioenergy facilities are located, especially ones using experimental technologies, exemplify the disconnect between the scales on which these discourses exist, circulate, and directly influence markets and behaviors. Here we show how various stakeholders spoke about different factors affecting potential threats and benefits in the context of the communities in which several different wood-based bioenergy facilities are located (mostly our primary sites in Georgia and Mississippi, though we draw from ethnographic work and in-depth interviews in other southeastern states). In doing so, we reveal the scalar and translational dynamics that become manifest when global discourses promote bioenergy development as a mechanism for mitigating global climate change and as a vehicle for economic resilience in small and rural communities traditionally dependent on forest resources, while local discourses are simultaneously circulating that often question climate science and resist government intervention in energy markets (e.g., in the form of subsidies for new bioenergy companies).

A key component of our research has been to explore how discourses and cultural models associated with bioenergy development affect how forest landowners and forestry professionals choose to manage forests in ways that might change forests if bioenergy development continues. The cultural understanding of trees and forests ultimately can and does shape forests themselves (Schelhas and Pfeffer 2005, 2009). These cultural understandings can be teased out of discourse (Quinn 2005); in the previous chapter, we showed numerous ways that ideas are circulating and being translated and contested among different opinion communities about bioenergy. Here, we look specifically at how people in our field sites talk about trees and forests in the context of extant or proposed wood-based bioenergy facilities located in their hometowns. We draw from the field of ethnoecology which focuses on how people classify, conceptualize, constitute and describe their environments, (see, for example, Anderson 2014, 24), which leads to understandings of how they establish, define, and negotiate strategies to manage natural resources.

Here we find it useful to return to the idea of the "triple bottom line" (see Simms 2003) of bioenergy development: ecological, economic, and social impacts on regions around new facilities that use wood as a primary feedstock. These impacts are deeply intertwined with each other and embedded within the complex social and political histories of the region as a whole and specific communities. We first explore the ways that people talk about forests and trees, linking their forest management behaviors and feelings about wood-based bioenergy to the core values that they associate with forests and trees. Then, using one specific case, Soperton, Georgia, we focus on the interplay between ideas about the ecological, economic, and social impacts of an experimental technology using an established commodity (pine trees); these complexities become evident when people in these communities refer to a number of different conventional discourses, such as government regulation, national security, and ecological sustainability.

WHAT PEOPLE TALK ABOUT WHEN THEY TALK ABOUT TREES

Valuations of Trees and Forests

People in the South love trees. This may be an oversimplification, but it's a statement that we stand by, living in the South ourselves and speaking with thousands of Southerners over the years, many of whom own or manage forest land. Different people love different trees, and they love trees in different ways. But the love for trees and forests (or, "the woods," according to most non-foresters) is evident even among the most utilitarian forestry professionals. We found multiple, complex, and sometimes incommensurable or competing values ascribed to southern forests, and our studies only begin to capture the complexities of how landowners value forests and the specific trees within them. Understanding how people feel about trees is absolutely essential in understanding their reactions to proposed energy developments relying on wood or the transformation of currently forested land.

Some people spoke of trees and forests in sentimental, almost romantic terms. They spoke of forests as places of comfort, safety, respite, and as places with intrinsic value. Forests were valued for more than timber, including wildlife, hunting, and aesthetics. One forest owner, a politically conservative White man who was actively engaged in growing pine trees for timber, expressed this notion particularly clearly:

> I love the forest for several reasons. Tree farming doesn't require the constant input necessary for row crops. I'm too old and don't have the technology and skills to do that. Forestry, I can do and be successful with it. . . . The other thing, I love having forest around me. It's sort of like a blanket. I can be next to God, go outside and walk the trails through the forest. Feel Mother Nature, see wildlife in the forest, the whole gamut. The forest is my blanket.

Other landowners talked about sentimental attachments to land because it's been in the family for generations. One landowner, a White man in Waycross, Georgia, said: "We're just trying not to lose the family land . . . . We have a legacy tract that's not for sale. Our granddad swapped a Model A for it." Family legacy was an important component of land ownership for many people with whom we spoke, and particularly so for Black Americans, whose forebears may have acquired land during Reconstruction, or sometimes even during times of slavery. In chapter 5, we focus on what land and forests, and especially family ownership of them, means to many of the Black landowners with whom we spoke; while there are certainly overlapping values among landowners of all races, the landscape has particularly deep cultural, historical, and social meanings for Black Americans.

This broad moral dimension in the ways that landowners talk about and relate to forests differs from the ways we observed people engaged in forest product and bioenergy industries talk about them. Specifically, industry participants tended to refer to wood obtained from forests as "feedstock" or "product," removing any association to land or forests. A forester in Mississippi told us: "The economy in Mississippi and Alabama . . . forests are the cash cow." Some community members in our field sites spoke of forests in more utilitarian terms, as resources and commodities. Many spoke of trees as an investment; one said: "timber is a better investment than the bank savings rate." A number also stated that while there are economic risks associated with planting trees to harvest (including tree death or damage from insects, diseases, and storms), the overall rate of return is higher and more secure than investing in the stock market. Many of these speakers were involved, directly or indirectly, in the forest products industry in the communities we studied. But again, the categories of people are overlapping, as are the valuations of forests.

What we found was that even the most industry-oriented people we interviewed cared about the long-term health of the region's forests. They might disagree about whether large-scale forest plantations are or are not fully functional ecosystems, and they might never agree on whether wood-based bioenergy is a net gain or loss for the southeastern landscape. As one landowner in Soperton said:

I am a true conservative. I love trees, but I don't hug trees. I'm very conservative. I think we should leave the planet better than we found it.

Sometimes people's ideas about trees changed over time. One forest landowner in Georgia told us that: "When I first owned timberland, it was primarily to make a profit. The older I got, the more I cared about the aesthetics."

Thoughts and feelings here about trees and forests are rooted in a long history of the forest products industry, which ultimately colors people's perceptions of new uses for wood, including how they valued different types of trees within these complex ecosystems. One way in which this becomes evident is when that landowners and forestry professionals talk about and ascribe meaning to hardwoods versus softwoods. For example, softwoods, especially pines, are seen as more economically productive (in appropriate ecological niches and with adequate acreage). Pines are heavily promoted by foresters and natural resource managers, and it is not uncommon for a landowner to choose to plant all pines in a monocrop plantation system by removing established hardwoods and preventing new understory growth. One forest industry employee in Waycross, Georgia, told us:

I would not plant a hardwood. I don't know why you would plant them, there's always something to do with pine. . . . I'd not plant any hardwood when we're surrounded by pines.

A few resource managers in Soperton, Georgia, echoed this sentiment, adding that new products are always being made out of pine trees, making it an eternally safe choice for new plantings. They said: "Pine trees are always so great. You think they're maxed out, then they find something new. Turpentine, sawtimber, paper." These particular people were also hopeful about using pine trees as biofuel, thereby supportive of the creation of new products from familiar sources.

However, not all landowners who have been steered to plant pines have been successful. The forestry maxim "plant your sorry, worn-out acres to trees" has caused much disappointment to ill-informed landowners who assume that pine trees grow well everywhere in the South and do not know to treat trees as they would other crops. One forest landowner in Columbus, Mississippi, said: "I got sucked into the pine tree deal. They said, it should be good land for pine." However, his property had a low site index,[1] meaning that pine trees would not grow to a height that would be profitable for him; he felt that professional foresters should have known better, but were too dedicated to pine trees to consider other alternatives. Another landowner told us that: "I hate to say this, but the quickest way to poverty is pine trees. There is a cash event only once every 16 years. The way to create wealth is mining or agriculture." This man was also a farmer, accustomed to receiving a profit each season.

Hardwoods are generally not as economically productive as pines, but they are still often considered valuable for wildlife, ecological function, and aesthetics. People are particularly fond of white oak (*Quercus alba*), which are large shade trees which produce a lot of food for deer, squirrels, and other wildlife. One woman said, at a meeting of a local garden club in Georgia:

> I love the hardwoods so much more than the pines. I really like the Cherokee dogwoods and the redbuds . . . . They shouldn't destroy the big oaks in the woods. They should go around them, be more selective when they do the logging.

She understands and accepts the need for logging, but she doesn't feel the economic benefit of monocropping pine trees is worth the loss of old oaks. At the pine-hardwood interface line in Mississippi, hardwoods are particularly popular for hunting (mostly deer). A landowner in Columbus, Mississippi, who was enthusiastic about deer hunting in his forests, happily showed us a wild black cherry (*Prunus serotina*) sapling on his property, saying: "This here is a pie tree. The big boys come for that pie." Many people with whom we spoke valued hardwood species simply for their beauty.

A minority of landowners, at least of those who grow trees to sell, prefer to grow hardwoods over softwoods; this is likely due to a combination of personal preference and local markets and site suitability for different species. A landowner in Columbus, Mississippi, who chose to grow hardwoods for profit stated: "People say hardwoods take too long. But you plant it like a garden, you water and fertilize them, and you can get a mass of them in seven or eight years. I'd always go with hardwoods." An extension agent in Mississippi said:

> If you want trees, choose the best. In specific soils, don't plant pine trees . . . Hardwoods are good for wildlife, and you can harvest way down the road. White oak, Nuttall, sawtooth, not live oak. Two months ago, hardwoods was the thing. It was selling good . . . ArborGen is working on hardwoods that you can harvest in twenty or thirty years.

A forestry extension professor in Mississippi, who works with a number of forest landowners, also said that some landowners prefer hardwoods. He said: "Some landowners, all they care about is hardwoods. They say, we don't care about pine. But leave our hardwoods alone. They want to keep the oaks." This is definitely a true statement, as a forest landowner told us in Mississippi: "To me a pine tree is worthless. I want the wildlife. It keeps America going. Most people won't plant hardwoods, 'cause they know they won't get nothing out of it in their lifetime." But as noted, some tree production companies are working on genetic alterations to produce harvestable hardwoods on a rotation that is competitive with pine, catering to both the preferences of some landowners and a range of soils less suitable for pine. Another large-scale landowner in Mississippi, who grows and harvests pine trees, talked about one special oak tree that he keeps, and he uses this example to demonstrate the way that some environmentalists misunderstand landowners' love for trees:

> There's an oak tree in this field. It costs me $1000 a year to keep it in the middle of the field. I won't cut it though. I plant 900 acres. Then those sons of bitches yell at me for cutting a tree, and they've never planted a tree in their lives.

So while there was clear preference for pine trees for forest industry professionals and landowners with large acreages who prioritize revenue from trees, this value was not universally shared, even in the heart of pine country. People here still have a soft spot in their hearts for hardwoods.

Hardwoods can also not be valued and, in some cases, considered "trash trees." The ones most often in this category included sweetgum (*Liquidambar styraciflua*), green ash (*Fraxinus pennsylvanica*), black locust (*Robinia pseudoacacia*), honey locust (*Gleditsia triacanthos*), osage orange (*Maclura pomifera*), and naturalized Bradford pears (*Pyrus calleryana* 'Bradford'). An extension agent in Mississippi told us: "Green ash is a nuisance. It's our wild growing tree . . . Sweetgum, we're all eat up with it." He also told us how much he hates green ash in particular, saying that it punctures tires, and that he's "hoping the Emerald Ash Borer makes it to Mississippi to kill all the ash." A landowner near Soperton, Georgia, said when showing us land near his: "Sweetgums suck all the water up. Trumpet vine strangles the trees. This area [has lots of] thick, trashy trees—it's good for deer and firewood, that's about it." One landowner in Columbus, Mississippi, told us that when planting a new stand, he plants both hardwoods and softwoods that he considers valuable but removes other species. He said:

> I'm a certified tree farmer. I have two 100-acre plots. They're pine and hardwoods. I took out the trash trees—gum, hackberry, ash. Anything that wasn't bearing. I left the native hardwoods there. Now I plant a mixture.

The hardwoods that he has planted include some American chestnuts and fifty acres of pecans.

Others, both landowners and forestry professionals, spoke of the importance of having a mixture of trees, even those broadly considered worthless or worse by the majority of forest landowners who are selling trees. A forester in Waycross, Georgia, pushed back against conventional pine monocropping by accepting the presence of some "trash trees" and preferring multi-species ecosystems:

> We don't do 100 % weed control. I don't like to just see 100 % pine trees. I like healthy forests. It's not healthy if it's only trees. I'm a forester, but I like to see quail. In a forest, it all has its place.

He elaborated this distinction as well: "I'm a forest manager, not a tree manager." One landowner in Soperton, Georgia, told us: "Many acres in Georgia are covered with undesirable forest. These are cutover forests that are regenerating in cherries, undesirable pines, and other junk." As noted, one landowner's junk is another's treasure; these trees, such as cherries, can often be useful in a biodiverse, multi-use landscape. Needless to say, environmental advocates see both intrinsic and ecological value in a diversity of plant and animal species and likely wouldn't refer to any native species as "trash," though they do have a number of concerns about invasive exotic species.

In addition to distinctions between softwoods, hardwoods, and "junk" or "trash," many people, mainly forest professionals and larger-scale forest landowners, have clear opinions about the value of different kinds of pine trees. These are also important to discuss here, since pine is the main biomass feedstock, and because these distinctions demonstrate the depth of familiarity that they have with the entire life cycle of a pine in the South, from pinecone to seedling to finished product, whether it's a telephone pole, a railroad tie, a

2″×4″ at Lowes, a roll of toilet paper, a wood pellet, or ethanol blended into gasoline at the pump. We focus on the three main commercial species in the rural South: loblolly pine (*Pinus taeda*), slash pine (*Pinus elliottii*), and long-leaf pine (*Pinus palustris*).

We found that some people are as loyal to specific pine species as they are to their favorite college football teams. Some prefer loblolly, as it is the most commonly planted pine in the region, due to its relatively short rotation, mul-tiple markets and end uses, and general hardiness; most soils in the region are also well-suited to this species. Some admitted no great love for the loblolly, but a strong faith in its ability to survive any changes in the markets. Other landowners greatly prefer slash pine. Others will only plant longleaf, which is taller, straighter, more disease and insect resistant and generally "pret-tier," but has a longer rotation and a more limited range. A natural resource manager in Columbus, Mississippi, told us that most of the CRP[2] land is in loblolly pine, because the best soils for longleaf are about eighty miles south of there. "They may grow," he said, "but not to production size."

Different values emerge when people talk about different types of trees. Several spoke of the versatility of different species. One landowner in Georgia told us that loblolly makes the best pulp, but boards made from it are not as strong as those made from slash pine, which he considered to be more useful for a variety of purposes. Slash and longleaf pine can both be used as sources of pinestraw for landscaping (though longleaf is considered the best). Pinestraw can be a lucrative business, especially as a source of income while trees grow to a harvestable size. One landowner in Waycross, Georgia, said: "I've got some loblolly, won't ever plant it again . . . I've got 150 acres of longleaf . . . The straw on the longleaf is worth more than the tree." One forestry professional in Georgia also cited the tension between aesthetics and practicality when he said:

> Loblolly makes pieces of shit. They're ugly. I don't like them. The limbs are all in your face when you're mowing underneath them. Slash bark is prettier. They're about the same height, but the dbh[3] is bigger in slash. I like longleaf better. But that's not a viable option for rotation. It takes too long.

In his estimation, shared by a number of people with whom we spoke, long-leaf is preferable when the land is suitable and if you're willing to wait. Slash pine is the next best choice, and loblolly will do otherwise.

While landowners, community members, and forestry professionals generally love to grow and be surrounded by trees, they mostly expressed concerns about nonnative species. In the context of bioenergy development that includes biomass from private forests, landowners generally appeared reluctant to grow exotic, genetically modified, and/or potentially invasive tree species such as eucalyptus (*Eucalyptus* spp.) or paulownia (*Paulownia* spp.) specifically for the bioenergy market. An extension forester in south-ern Georgia said of the landowners he works with: "You can count on one hand the ones that are willing to grow something other than pine . . . with eucalyptus, there are no other options like with pine." This reluctance to introduce new species give clues as to how bioenergy development may—or may not—lead to shifts in valuations of Southern forests and trees, given how much land is privately managed in the southeastern United States. However, we did note that some landowners and forestry professionals were open to the idea of growing exotic species, believing that despite concerns, they are still a better alternative to fossil fuels. One forest landowner in Waycross, Georgia, said when asked if he and his family would be willing to grow and manage eucalyptus for a bioenergy market:

> Yes, we really would. We'd be interested in looking at anything. Look, I grew up a hippie. I like the idea of sustainable resources. I'd like to see us get away from coal, and I see greater potential for timber being a valuable resource [for energy].

He was particularly excited about the potential for renewable energy from locally grown resources such as woody biomass, even from nonlocal species to reduce or replace the needs for fracking and the extraction of coal.

WOOD-BASED BIOENERGY IN FOREST-DEPENDENT COMMUNITIES

The forest products industry has always been a key component of the economy of the southeastern United States, and bioenergy is promoted in the region as a market for surplus wood. However, contrasting narratives about impacts of woody biomass development on forested landscapes circulate among representatives of forest products industries and forest landowners who supply them. Some forest products companies welcome woody biomass as a new market, while others feel threatened by it. Sarah Mittlefehldt (2016, 18) notes

> a kind of narrative inconsistency among the forest products industry: some felt that the burgeoning fuelwood market could help bolster other forest products, while others believed that traditional forest product markets might be threatened if the demand for forest fuels increased significantly.

Most foresters, natural resource professionals, forest products industry representatives (including wood-based bioenergy producers) with whom we spoke supported the idea that a new market for wood was a good thing for the forest industry as a whole because it provided local jobs, income for landowners, and incentives to maintain forested acres. A state agency forester told us, "There's more wood there than is used by pulp mills, small diameter stuff anyway." Wood product markets are seen as critical for retention and management of the forest plantation systems in the South. As one forester said at a workshop, "the biggest threat to forests is no markets. Wood for energy is a key for forestry in the Southern U.S. . . . Healthy forests need markets for forest products, and healthy forests need markets for management by-products."

However, we did also speak with some owners of small-scale forest products companies who felt that demand for woody biomass from large-scale wood-based bioenergy facilities would put too great a strain on local resources and also directly threaten the survival of their companies. The owner of a mulching company in southern Georgia told us, "We do mulch here, so there's direct competition that that stuff . . . The pellet mill is propped up with federal and state money. We don't have anyone helping us."

Previous studies have shown that while there is generally broad public support for renewable energy, when actual implantation begins and the impacts become apparent, community members are more likely to resist (Bailey and Darkal 2018; Bidwell 2013; Wolsink 2007). Berlik et al. (2002) note that in places where energy consumption is particularly high, energy production is often low, even though affluent countries are better equipped to extract resources with minimal environmental damage. When the externalities of energy consumption are hidden from view, the tendency for many communities is to protect their own resources and viewscapes while exploiting those out of sight. Mittlefehldt (2016, 17) notes that since the 1970s, forest professionals have expended a lot of energy "to convince the public that clearcutting was an ecologically sound practice and was sometimes necessary in order to regenerate certain early successional tree species."

In some areas, particularly those not accustomed to regular, large-scale tree harvesting, the proliferation of wood-based bioenergy has led to opposition from those concerned about ecological fragmentation or about the alteration of scenic landscapes. It has also, in some cases, "forced communities to see the consequences of their energy consumption embedded in forest landscapes" (Mittlefehldt 2016, 17). Berlick et al. 2002 ask the question: what if communities saw the effects of their own energy consumption? In other words, what does local energy independence actually look like? It would look different in different places of course, but in their study in Massachusetts, where energy consumption is high, as is support for environmental regulations limiting harvesting, they showed that it is nearly impossible to meet current energy demand with biomass from sustainably managed forests. However, other alternative routes to sustainability exist, including reducing wood consumption, reducing energy consumption, increasing wood recycling, conserving forested environments and protecting fragile ecosystems, and encouraging sustainable wood production in areas that are the most suitable. These same recommendations would apply in the southeastern United States as well, though it is important to note the effects of long history of engagement with forestry in this region.

Members of historically forest-dependent communities and those involved in wood products industries have a pragmatic approach to the harvesting of trees for multiple uses, and they are not as concerned about the aesthetics of clear-cutting or the effects of traffic congestion caused by transporting logs or logging equipment as are members of the general public (figure 3.1). In the U.S. South, where many people understand forestry as a cyclical event involving a renewable resource, local production of energy from trees, for many people, would be a source of pride rather than horror. People that we met from communities with wood-based bioenergy plants were widely engaged with and accustomed to all aspects of forestry and the forest products industry. They generally believed there were plenty of trees in their region, noting the compatibility of a new market for wood products in rural areas with many pine forests. One person said, "This is a woodbasket, a heavily forested area. There's a surplus because of the loss of a lot of pulp mills." Another commented: "Here people are used to seeing logging trucks, trains. It doesn't freak people out to see trees cut." They believed

Figure 3.1   Pine Logs Harvested for Wood-based Bioenergy in Southern Georgia, 2011.
Source: Credit: Sarah Hitchner.

that biomass harvests would be good for forest cover, noting that adding new markets provides incentives for landowners to plant more trees. Other community members warned of potential negative consequences for forest health if surplus wood is not used. A functioning planted pine ecosystem, they explained, was dependent on sustainable forest management practices that include both planting and cutting at scheduled intervals.

At the same time, landowners had little interest in changing the ways they grow trees. The existing system of plantations of native pine, in which trees are thinned for pulp in the first two decades and then grown into sawn timber or poles, was seen as flexible and durable. People believed there have always been markets for pine trees over the past century and were resistant to large changes in the system. They saw harvests for biomass fitting into this system, but not wholly replacing it with trees grown only for biomass, because the final products, sawtimber and poles, were the most valuable part of the system and made the system work economically. They did not see production of bioenergy feedstocks as the sole goal of forest management. One person stated, "forest landowners may switch to a shorter rotation, but sawtimber is still the goal." A forest landowner said:

> I'm all for it [bioenergy]. But I wouldn't grow just for that. I don't think it will ever be an industry in itself. People need to have multiple products—both landowners and companies—you can't put all your eggs in one basket.

One forester noted that "bioenergy works best when integrated with other forest products." In many rural communities in the U.S. South, there is a long tradition of forest management for timber and other wood-based products; the integration of bioenergy development into these existing management frameworks, rather than competition with them, is often seen as key to the regional success of bioenergy development (Karlsson and Wolf 2008; Wolf et al. 2006).

Prefiguring the cycle of disillusionment we described earlier, the biggest point of contention from communities and forest owners was that every bioenergy plant came into town saying that they were going to buy residues and waste wood for which there was not currently a market. This would include logging residues and trees smaller than those used for pulpwood, a resource that exists but may not be economically harvestable (Perlack et al. 2005; Jeuck and Duncan 2009). In the development stage, using a nonutilized waste product both fit the larger imaginary and the promised greater economic return to forest owners. This practice would provide both direct revenue for products that previously had no economic value and also reduced costs for site preparation for new plantings, which involves removing waste products.

One person explained that landowners were "excited about the possibility of having a market for trash left over from harvests . . . limbs, tops, stumps, to clean up the land."

However, when markets for early thinnings and residues did not materialize after the bioenergy plants began production, there was widespread disillusionment about the preliminary statements by bioenergy companies that they would buy these materials. One landowner said they believed the facility "would buy our leftover products from cutting timber. That was wishful thinking maybe on our part." Another said: "They were blowing smoke . . . everyone was excited about them taking thinnings." Another also expressed anger in discussing how the companies promised to take slash piles left over after a timber harvest, naming a specific forester they felt was being deliberately dishonest:

> They said they'd take everything. [*Forester's name*] told us all that bullshit. In the end, they're only taking clean chips. At their first presentation, they said they'd take leaves and scraps. That gets people excited.

Despite promises to the contrary, all the plants ended up buying the equivalent of pulpwood. A wood buyer at a pellet company said, "We buy the same thing as the pulp mills. We buy pine pulpwood. Two-inch tops, limbed. . . . We call it pine fiber. . . . It's cheaper to buy roundwood and pulpwood than waste." Similarly, while forest owners hoped that bioenergy development would lead to higher wood prices, many forest owners felt that there has been little effect on wood markets. However, professional foresters took a more optimistic view, saying that it has stabilized pulp prices or kept them from dropping lower.

While some industrial developments (e.g., Range Fuels in Soperton, GA) planned to use waste wood that is currently left on-site, making this economically feasible is a challenge. The economics of wood transport mean that harvesting will be concentrated near biofuel plants. We found that if it is profitable to have biomass as a production goal, many landowners (but by no means all) are willing to shorten the rotation on some of their forest tracts. At the same time, sawtimber has historically been the product with greatest returns, and many forest landowners will continue to manage tracts for it (but will allow thinning for biomass during early years), as well as for hunting and recreation. As noted, while there are experiments with fast-growing tree species targeted for bioenergy (e.g., eucalyptus), most landowners prefer traditional forestry species (loblolly, longleaf, and slash pine) that they perceive to be less risky because of existing markets and multiple uses. In some study sites (e.g., near Soperton), energy grasses are being planted, and there are indications that, if markets exist, some landowners may convert poorer forestry sites to grasses. However, given the failure of the Range Fuels facility there, this seems even less likely to be an attractive option for the vast majority of private landowners.

In discussing the sometimes incommensurable perceptions by various community members of the ecological, economic, and cultural impacts of bioenergy development on local forest-dependent communities in the southeastern United States, we found that many opinions are heavily couched in political ideology. We found that political affiliation and ideas about the role of the federal and state government in local development options permeated all of these discussions, and none can be interpreted independently of politics. We found this to be especially true when landowners and forestry professionals spoke about government interventions into forest management such as certification, a public marker of renewability and sustainability, energy security, government subsidies for renewable energy, and even within intra-community dynamics affected by local bioenergy developments.

Renewability and Sustainability

If bioenergy development is to meet the expectation of the national imaginary, it must be seen as renewable and sustainable. This includes reduction of greenhouse gas emissions to address climate change and sustainable production of biomass feedstocks. International and national discussions of bioenergy and government policies that promote it emphasize the need for these to be measured and certified. Many actors involved in promoting bioenergy in the southeastern United States recognized that the concerns raised by the counternarratives about renewability and threats to forest ecosystems are real, or at least have powerful constituencies.

Brown has described the word "sustainability" as an "empty signifier" which is, therefore, "subject to radically diverse interpretations" (Brown 2015, 116). Renewability and sustainability have long been claimed by forestry and forest owners, although their interpretation of these terms tends to have a narrow focus. These terms may refer only to the continued production of wood products, be expanded to include wildlife (both game and nongame species), or more broadly approach national definitions. We found that people in the southeastern United States assume that forestry is a sustainable way to grow a renewable resource. As one forester noted, "Trees are truly renewable. The forest products industry has been doing sustainable forest management for a long time." Similarly, a forest owner noted that trees are a renewable resource, and that he would like to see a bioenergy market develop so that more of each tree is utilized; he said, "We hope that we can use the whole tree. There is so much waste when all we can do is burn it or let it rot." Showing the extent to which an assumption of renewability and sustainability has been internalized at the local level, one person associated with community development tried to extend sustainability to include the social dimension: "We're making a sustainable, renewable product, and we want all the processes, including the workforce, to be sustainable and renewable." These quotes demonstrate the almost ubiquitous belief that sustainable forest management is important and that people in this region are knowledgeable about how to do it.

Along with the tendency to view forest management as inherently sustainable and trees as a renewable resource, there was resistance to forest certification. The tendency was for people to see it as outside interference in their forest management, which they believed was already sustainable. As an industry forester said, "[Certification] is for what producers already do." Another forester said:

> They [landowners] don't like being told what to do. They don't like that with certification, third parties have the right to audit. . . . It costs us $2500 a year to have someone tell us we're doing it right . . . . It's going to be a hard sell. People don't want any new regulations shoved down their throat.

One forester was openly hostile to forest certification, linking it to environmental and social agendas: "Certification is a tool for people with an agenda . . . a political agenda." Another suggested that even when required, it wasn't taken seriously: "[*Company name*] requires SFI, to cover their ass. They don't really check it. But it's okay, it's good." Similarly, another forester said: "Certification is only a marketing tool. It's hogwash." But then he went on to admit that it may be needed: "The public perception of certification is that it's more sustainable. You have to acknowledge that." Ultimately, those involved in the forest products industry tended to see certification as inevitable: "Landowners also will need to demonstrate sustainability. But they don't want to do this. . . . Landowners need to unite and encourage the markets to come to them. If they don't, the industry will pass them by."

Energy Security

As discussed in the previous chapter, energy independence and security were key elements of the national bioenergy imaginary and related discourse; these values were also expressed on the ground in our field sites. A Georgia state employee promoting bioenergy development said, when speaking about liquid fuels made from woody biomass: "Look at the Middle East nightmare. We need it." Speakers at local workshops further referenced the hostility of people in this region to the United States; for example, one said, "Energy independence is high on my list. Every gallon [we] produce is one gallon we're not importing from places where they don't like us very much." At a local workshop promoting bioenergy feedstock development, several speakers explicitly noted the link between dependence on foreign oil and threat of harm to U.S. citizens at home and abroad, especially to members of the military stationed in these oil-rich and politically unstable areas. One speaker noted that in order to "get this oil crisis solved," it was necessary to "connect the U.S. energy policies and the dead troops. All Congressmen should see all those bodies at Andrews Air Force Base." Noting bioenergy as a potential solution, another said, "I have no desire to send our young men and women to fight so we can drive SUVs and pickups."

Reducing dependence on foreign oil and energy independence were powerful justifications in local discourse. One forester said: "I'd love for us not to be so dependent on foreign oil. . . . I'd like to see it work. Be independent of Saudi Arabia and Iran and all that crap." It can be tempered, however, by concern for gas prices and skepticism regarding the economics of new bioenergy development and how bioenergy policy might directly affect prices of energy products such as gasoline for consumers, as this quote from a community official indicates: "Yes, it's great to reduce our dependence on foreign oil, but it comes down to ye old pocketbook." More commonly, local discourses endorse energy independence and describe the need for a sustained effort for new technology development in the context of the new industries that are coming to their towns. One local development official explained that people need to understand the global power dynamics regarding energy, despite the domestic political power of the oil and gas lobby in the United States:

> The Middle East has the oil, and they have a lot of power over the U.S. They have us hostage. When renewable energy makes progress in the U.S., they drop the price of oil, and gas prices drop. They're smart, but we can outsmart them. But I don't know if we have the gumption to do it . . . Congress is a bunch of do-nothings. The oil and gas lobby just has to whimper, and Congress asks: "what do you need?" . . . People are not aware that the U.S. could face a real dilemma if we offend the Middle Eastern countries. If they cut off oil to us, it would put us back into the '20s and '30s. I'm exaggerating, but I'm exaggerating to make a point. People don't realize that we don't have a Plan B, other than bioenergy.

This viewpoint reiterates the opinion that energy independence is not just desirable but indeed a matter of national survival. To counter the power that oil and gas lobbies have over policy-makers, bioenergy proponents must produce a strong argument that bioenergy development deserves at least as much attention and funding as fossil fuel industries. A key component of this argument relates to the potential economic development that bioenergy can bring to rural areas. This is especially true for communities with abundant, merchantable trees and a long history of engagement with forest product industries (which includes an acceptance of tree harvest and a workforce with forestry experience).

Rural Development; or Intracommunity Dynamics

Rural development was frequently highlighted by bioenergy promoters in local communities, knowing that it would be a selling point helping to overcome any potential reluctance about inviting new companies and technologies to rural areas. They also often referenced "idle lands" or land that they felt could be put to its "greatest and highest use" by planting woody biomass for energy feedstocks. At a workshop promoting bioenergy feedstock development, a speaker said: "We would see prosperity return to rural America. Let's get all those idle acres planted." At a meeting of a local association, a speaker from KiOR said:

> We're about rural America. We come to places like Columbus and take feedstocks and convert them to fuel. . . . Everywhere KiOR can build a plant, it can stimulate the local economy.

Correspondingly, the need for rural development was widely acknowledged by people in towns that have seen local businesses and jobs decline as trees have replaced farms and transportation improvements have shifted jobs and shopping opportunities to larger regional centers. Local people talk about closings of wood products mills and factories and their pursuit of any development that would bring jobs, ranging from manufacturing to prisons. As one community member said: "The only jobs are at the school. People work, bank, shop, and eat out of town. It's difficult to get a long-term employer even though we have a good location near the interstate with good roads." The desire for development was so strong that several people noted that they would take anything short of hazardous waste. In one community, someone said: "This area will take almost any kind of development. The only thing people here fought was a large waste facility on the river." Similarly, a prominent community member in another Southern town said: "Development would be good, but not something like a hazardous waste dump." Local community members are desperate for any development that does not threaten public safety; bioenergy, for the most part, fulfills this requirement.

However, experiences with bioenergy development in communities where it took place on the ground inspired both hope and disappointment for improved rural development. A community leader in one town said: "My vision for our town is more economic development, a diversified job market—high tech if possible." One county official in Georgia noted that a new forest products industry would be very helpful for the county because it could match available resources with employment opportunities; he stated: "We desperately need jobs here; there's a lot of wood and a lot of unemployed people." A state official involved in matching bioenergy development with communities described his experience at a public meeting related to a bio-energy plant: "People were eager for markets. The hard part was being able to leave—people just kept wanting to talk. They were excited about rural development." Both community members and landowners expressed interest in increased job opportunities and added income for farmers and forest own-ers. After plants were built, however, many local people complained that the local bioenergy development failed to create the jobs that the companies had promised. Community members frequently complained that in addition to the fact that fewer jobs than expected were created, local people were often not hired in the jobs that did become available. We discuss this in detail in the following section, using Range Fuels in Soperton, Georgia, as an example.

Similar complaints were raised about the contribution of the bioenergy plants to wood markets, as well as the ways that increases in these markets would benefit some community members and not others. Community mem-bers without large landholdings felt that large landowners, who were also often community leaders, were promoting bioenergy as a means to obtain personal gain. This exacerbated underlying class tensions within communi-ties. One forest owner without much acreage said, "I mean, everyone would benefit from bioenergy, but not to the extent of the big landowners. They can sell directly to the plant." However, it's worth noting that the perceived benefits to large landowners did not materialize, as the wood markets did not change enough to benefit landowners in the ways that companies had prom-ised. As one forester said, "Some people are not happy with the way they came in and presented themselves. It's true they brought some jobs. But they didn't help with thinnings. It was misleading."

In addition to the perception of unequal benefits to community members based on landholdings, we also noted perceived racial biases in the towns which led community members to make direct statements about the inequi-table distribution of benefits from bioenergy development, particularly with regard to high-paying jobs at bioenergy facilities. We explore this and other environmental justice issues in chapter 5.

Government Promotion and Investment

Bioenergy promoters often argued that technologies like pellets and even cellulosic fuels are just the beginning in the development of a large, sophis-ticated bio-based industry in the southeastern United States. A state official described pellet plants as laying the groundwork for a future liquid fuels industry. Commonly cited future products include drop-in fuels, cellulosic ethanol, chemicals, and pharmaceuticals. Improvements in biomass-processing technologies such as torrefaction, fermentation, and gasification were described as leading to new products and more efficient production of marketable bio-based products. While some of these technologies are still in the experimental phase, local community leaders and members are calculating the potential risks and benefits along their "triple bottom line" as they assess potential ecological, economic, and social impacts of bioenergy development in their communities.

Local people also reacted in different ways to the fact that bioenergy plants came to their towns through government promotional efforts that included subsidies. Some, given setbacks in plant development and after plant failures, referred to communities being "scammed," "hoodwinked," or being sold "snake oil." A former mayor of a town in Mississippi noted that the town had provided a lot of money for a beef processing plant that went bankrupt. He said that as a result, people here "have a healthy dose of skepticism" of government investment in private businesses and tend to view government subsidized bioenergy as a slick sham that only some people know how to access, stating that "They wonder if KiOR is another beef plant."

Other people, particularly those involved in promoting development, countered these arguments by suggesting that other types of energy, and many other industries in general, receive subsidies. They suggested that bioenergy needs subsidies in order to compete with these other subsidized energy products, or that the subsidies bioenergy got are small in comparison to other industries. One person stated, for example: "It's hard to compete with coal; it's not a level field because coal is subsidized. . . . Other things get subsidies . . . oil, natural gas. . . There's nothing for biofuels." Several other interviewees in local communities mentioned the narrow mindset regarding opposition to subsidies for bioenergy development. Another person, noting that a lot of people are skeptical about government investment in renewable energy, made the point that they are directly benefiting from governmental investments in what are now everyday technologies; he asked: "What about government investment in cars, planes? They think that all happened without help from the government? Roads? Hello?" Others emphasized the hypocrisy of some people opposed to government subsidies and government intervention in markets in general. One person stated:

> Sure, there's subsidies. Farmer subsidies, commodity subsidies. There's insurance—farmers get money for spoiled crops . . . . There's this disconnect—people are anti-government but they want their subsidies. They go off on people on welfare and the welfare system, but by god, they want the subsidies coming to them.

The question of subsidizing new and emerging technologies, and the government's role in technological innovation in general, was a divisive one in all three of our primary field sites. We found that opinions ranged widely, but individuals more involved in promoting economic development or expecting to be employed or sell wood to the facility were generally more positive than people not directly involved, particularly minorities and poorer community members. Opinions on government investment in bioenergy development were influenced by political affiliation, with more conservative community members tending to be more skeptical about or resistant to subsidies for renewable energy developments such as wood-based bioenergy.

BIOENERGY DEVELOPMENT ON THE GROUND: THE CASE OF SOPERTON, GEORGIA

People on the ground in these rural, forest-dependent communities in which bioenergy development was happening at the time of our research had a lot to say about the potential impacts on their forests, local economies, and social dynamics. The ideas they circulated among themselves and expressed to us in interviews or casual conversation were, as they often told us, influenced by discussions they'd had with other community members, landowners, local loggers or foresters, and, in some cases, with bioenergy industry representatives. Their ideas about using trees as a source of energy are also rooted not just in political ideology, but also in their own sentimental attachments to land, forests, and even particular trees. In order to understand what people think about wood-based bioenergy, it is necessary to first understand their fundamental relationship to forested ecosystems and to the forest product industry. As noted, most of the people in the communities with whom we spoke understand the long history of forest products in their area and welcome the opportunity for trees to again become a source of employment and wealth. How that does or does not actually happen in particular sites gets messy. Here we explore the messiness in one small town in Georgia.

As noted, Soperton, Georgia, is the site of what was going to be world's first commercial-scale cellulosic ethanol production facility, Range Fuels. We've opened this book with a spoiler: it didn't work. Here we will place this failure in context in this rural community with a long history of active engagement with the forest industry. We discuss the perceptions of Range Fuels during the planning stages (and brief operational stage) as well as the reaction of different stakeholders when it failed and was then bought by LanzaTech, another company that also produces biofuels (among other products). Our research in and around Soperton, Georgia, exemplifies the cycle of enthusiasm and disillusionment, showing how the sociotechnical imaginary that promotes bioenergy as a driver of community resilience does not capture the risks and consequences of the potential failure of a bioenergy facility in a small, forest-dependent town and how it directly competes with a political imaginary in which government and business entities profit at the expense of communities and taxpayers regardless of whether a bioenergy project succeeds or fails. This case illustrates the complications that arise when multiple imaginaries—often multi-scalar and incommensurable—meet on the ground in specific locations.

Soperton, Georgia: The Million Pines City

Soperton, the Million Pines City, has long been a forest-dependent community, with an economy built on turpentine, naval stores, pulp, whole logs, and lumber. It claims to be the birthplace of the silvicultural practice of intentionally planting pines. Beginning in 1926, a farmer named James Fowler was one of the first people (some people, especially in Soperton, say the first person) to plant pine trees as a crop (he planted more than 7 million slash pines on several thousand acres in the area); before this, trees in the U.S. South were harvested and forests were allowed to naturally regenerate. The town of Soperton adopted its nickname in honor of Fowler's Million Pines Plantation, and each year the town hosts the Million Pines Arts and Crafts Festival. While scientific experiments with pine plantations and silviculture were beginning throughout the South in the early twentieth century (Fox et al. 2004), Fowler worked with Charles Herty's nearby Savannah Pulp and Paper Laboratory, which pioneered the modern pulp and paper industry and created a market for planation pines (Cooksey 2018). The 1933 printing of the *Soperton News*, a weekly publication, was the first newspaper to be printed on paper made from the pulp of pine trees, and that pulp came from Mr. Fowler's trees. As noted in our introduction, a copy of this first edition is now housed in the Smithsonian Institute in Washington, DC, a point of great pride to people that live there now. Several people specifically linked forest-based bioenergy development in Soperton to its identity as the Million Pines City. A community leader told us: "Biofuels are a good opportunity for Treutlen County. Soperton is the Million Pines City." Another community leader relayed to us the town's history as a way forward for the community:

> We'd love to see Treutlen County as the capital of bioenergy. We've been known as the Million Pines community for so long. If they produce jet fuel, Treutlen County will be known as the granddaddy of all places . . . In 1937, Jim Fowler, working with Dr. Herty, used pines from Treutlen County to make the

first paper (from trees). LanzaTech could put their chip operation on Fowler's property, and we could play on that. It's a unique situation evolving here.

As a pioneer in southern forestry tree planting and development of pulp for new uses, it seemed logical to people in Soperton that they would be first in the next big advancement in forestry.

The nickname of "Million Pines City" is rather an overstatement. Not the first part, as there are likely millions of pines in the county. But "city" is a stretch, as the town limits of Soperton include just over three square miles, and its population hovers at just over 3,000. Built at the crossroads of railroad lines, as many southern towns were, it was once much more prosperous and economically and socially vibrant than it is now. Today, like so many towns in the rural South, buildings downtown sit empty, boarded up, desolate and deteriorating, foot traffic a thing of the past. A number of interviewees expressed sadness about the economic decline of the town and the county. Some described in detail the times in the past when the area was thriving and then how businesses dried up and young people went away. This history of outmigration is common in small towns throughout the Southeast, and the lack of industry only exacerbates the problem. A forest landowner told us:

> This is an impoverished county. We have nothing. No manufacturing, just a little ag [agriculture] . . . [This area] was so different when I was a child. There were two car dealers, a movie theater, a tractor dealership. Now we just shake our heads . . . I just wonder what has happened to this little community.

Another forest landowner explained why he thought the vibrancy of earlier days was now gone:

> With good roads and improved gas mileage, the little towns like Soperton dried up. Towns are getting farther apart from each other . . . There was a Black migration north in the 1950s. There was integration and consolidation. Industry moved north too.

When the main industries left town (there were several textile mills that employed a large number of people, in addition to the thriving forest products industry), nothing replaced them. A forest industry professional said that: "The Industrial Age passed us by here . . . There's no financial support [in the county] except the land base." Without other industries, new companies are usually unwilling to relocate or open a new location in a rural area like this, no matter how large and skilled (or trainable) a potential workforce is. As one community leader told us, it becomes a chicken and egg situation:

We need jobs. In the past, we could have attracted industry. Now people ask, "Why don't they have industry here?" It's a gamble to come where there's no industry.

Because few new people move into the community, and because young people continue to leave and not come back, business owners close shop with little incentive to gamble on opening a new one. One person simply said: "Soperton's a cemetery now. This town died." Most interviewees expressed great pride in the history of their town, as well as the desire for more industry to come to the area, which would help stem outmigration and promote economic revival. Bioenergy seemed to hold a lot of promise on many fronts.

Range Fuels: When a Plant Comes to Town

And so the cycle begins with the production of enthusiasm. It is difficult to overstate the excitement in the bioenergy world about Range Fuels scaling up its gasification technology (which had been successful at its pilot plant in Colorado) to commercial scale in Soperton, Georgia. Commercial-scale production of cellulosic ethanol made from trees, mostly forest residues or woody materials left behind after logging operations, was going to be a breakthrough technology, revolutionize the energy sector, create new markets for products that people wanted to get rid of anyway, and again position Soperton as a leader in forest-based innovations. Everyone in the energy sector was paying close attention to Soperton; it was a huge opportunity for the bioenergy industry, government agencies promoting renewable energies, and landowners and community members in Soperton.

Construction began on Range Fuels in Soperton in November 2007, after the company received a US$76 million grant from the U.S. Department of Energy, followed in 2009 by an US$80 million loan guarantee from U.S. Department of Agriculture Rural Development through the Biorefinery Assistance Program authorized by the 2008 Food, Conservation, and Energy Act (although it did not actually receive all these funds). The company was also backed financially by more than US$160 million of investor funds and a US$6.25 million grant from the state of Georgia (Lane 2011), as well as a US$80 million construction loan from AgSouth Farm Credit bank, on which it defaulted (Lane 2011). Range Fuels was expected to produce 40 million gallons per year of cellulosic ethanol using southern yellow pine[4] as a feedstock and gasification technology developed in its demonstration scale plant in Colorado. As we described, the Range Fuels groundbreaking was attended by high-ranking government officials, including the U.S. secretary of energy.

In Soperton, the initial announcement of the plant was met with great enthusiasm, as it would bring many jobs to Treutlen County as well as a new market for wood products, including forest residue that previously had little or no monetary value. Several people talked about the optimism they or other community members felt when they learned that Range Fuels would be located in Soperton, both as a link to their history in forestry and as a driver for future economic development. It would be, said one community leader, "an opportunity to put Treutlen County on the map." Those people that owned or managed forestland were particularly enthused; as one forester told us: "We were pumped when it [Range Fuels] was coming. We had monthly meetings, went to the plant, wore a hardhat . . . We'd take anything as long as they use wood to do it." Many interviewees had forestland and view it as an investment, an icon of community identity, and a common thread linking the past and the future through steady, dependable income that does not require a full-time commitment. Forest landowners were optimistic that a new market for woody biomass would drive up the prices for forest products: A forest landowner and farmer said: "We need biomass to drive the prices up." Another forest landowner told us: "I'm for it. I know I'm biased. I grow trees for a living . . . I hope it may increase prices, make more competition."

Other community members were more motivated to support the new company because they saw it as a potential place of employment. One community member was frank in his assessment of the "enthusiasm" about Range Fuels that he witnessed:

> That was the big thing, because of job opportunities. People talked about it. The concerns at first were just: Who got hired? Were they local or not? People only care about money and getting employment.

As we shall see, concerns about the actual employment opportunities offered to local community members were justified. This optimism about jobs for locals, an influx of money into the city and county from spin-off businesses, higher prices for landowners' wood products, and good publicity for Soperton fits into the imaginary promulgated by national and international actors where bioenergy would help small towns like Soperton and also contribute to the goals of energy independence and climate change mitigation.

The cycle shifted toward disillusionment when this optimism dissipated as it became clear that Range Fuels would provide neither many local jobs nor higher prices for wood products and new markets for wood waste products. Range Fuels hired very few local people, and people were particularly upset about this. A number of community members talked about the broken promises regarding employment for local people; they said that Range Fuels instead hired people from nearby (bigger) towns, such as the town of Dublin (a larger town about twenty-five miles away). A community leader told us: "Range Fuels hired people from Dublin that were working in the paper company there. This was seen as jumping over Treutlen County folks." In some cases, Range Fuels might have required expertise not available in Soperton, but the perception was that for many of the available jobs, there were qualified people within the community that were not hired. One forest landowner said, "They kept saying there would be jobs. But they brought in people from out of town, even the construction workers." Another forest landowner said that while they expected to see more local employment, Range Fuels instead "imported people from Michigan," which he says "was not right." The reason for these hirings from pools of people from outside the community were not necessarily a direct result of decisions made by Range Fuels employees, but rather a consequence of who Range Fuels put in charge of hiring on the ground. One community leader explained it this way:

> Range Fuels didn't hire local people, though one of their stated purposes was to employ local people . . . Y'all done us wrong. There are qualified mill workers, but it's a buddy-buddy system. The guy doing the hiring . . . he hired his buddies.

This "buddy-buddy system," also referred to as the "good old boy network" is no new phenomenon in rural towns such as this one. It is an extension of a well-established system of White privilege that rarely extends to Black, or other minority, workers. We explore how this racial dynamic plays out in more detail in chapter 5.

Another source of disillusionment for many community members was what they perceived as the broken promise of Range Fuels to use waste wood that is currently left on-site (tops and branches). Forest landowners and forestry professionals questioned this assertion from the beginning, knowing the challenges of making this economically feasible. While landowner enthusiasm for this is high, as it would bring profits from previously unmerchantable material, in-woods chipping of forest residues is expensive, especially on smaller tracts. One forester's in-woods chipping experiment near Soperton found that the chips would have to sell for US$90 a ton (several times the going rate); he was more enthused about whole tree harvesting and transport. This, however, does not fit into the imaginary that Range Fuels was promoting—that they would use waste products to create fuel. Using whole trees for fuel would also be a lightning rod for environmentalists, something that Range Fuels was not keen to be. They wanted to be seen as a win-win-win solution for rural development, energy production from local sources, and recycling of waste.

In any case, forest landowners expected that the location of Range Fuels in Soperton would drive up wood prices, thereby increasing income for local forest landowners. This did not turn out to be the case. A forest products industry employee told us that: "We thought prices may go up, but they haven't. The mill just pays what they want to pay. There's too much pine and not enough demand." Oversupply is the main issue that allows mills to set prices; this makes it hard for forest owners to profit, even when transportation costs are reduced due to closer mill proximity. One forest landowner explained this cycle of enthusiasm and disappointment in regard to potential for increased profits for forest landowners:

> People were optimistic, but not now. They were excited, thought it would help pulp prices if it was less distance to the mill. People were really enthusiastic about a mill here . . . that it would help stumpage prices.

These multiple failures to meet community and landowner expectations while the plant was constructed and operational further soured public opinion on bioenergy development in general. When the plant then closed, opinions about bioenergy sunk even lower.

Owing to technical problems in scaling up the technology, Range Fuels was only able to produce one batch of methanol from synthesis gas, which was possible in the 1920s and is a common process today (Lane 2011). In 2011, Range Fuels declared bankruptcy and closed. The implications of the failure of Range Fuels have been profound, both within the national biofuel industry and within the communities in and around Soperton. Because the opening of the facility was so heavily publicized and attended by high-profile people who seemed to be able to accurately gauge its potential, there was a lot of expectation that it would be successful. One forester told us: "There was a lot of hoopla about it [Range Fuels], that it would be the salvation of the county. The secretary of energy, the governor, they were all here." When Range Fuels did close a short time later, community members felt a range of emotions, including disappointment, anger, and disillusionment. Some people discussed a sense of disappointment and sadness, given that they felt that a development like this was really the only hope that the town had for an economic and social revival:

> We were most, most disappointed. We had such high hopes, because we have so little . . . A lot of people were . . . "enraged" is too strong a word . . . But people had high hopes that it would provide jobs for the community.
>
> We were let down when it didn't go. It would have been an asset to the community . . . Some people have a negative attitude about everything. But most people thought it would be an asset. It was our only option.

Some talked specifically about how this failure led to deeper skepticism about bioenergy in general. Several forest landowners told us that it was such a painful experience for the community and such a source of disillusionment that people rarely even talk about bioenergy anymore:

> I haven't heard much talk about it [bioenergy] recently. There was lots of talk. People were very excited, upbeat. Then you see something fail, you get skeptical. Now they're like "we'll just wait and see if it works. I'll believe it when I see it.
>
> We were all excited about Range Fuels . . . They just didn't do their research . . . People were so disheartened [when Range Fuels closed]. No one talks about it now . . . It just bust their heart.

Other people also mentioned the high hopes in the community following the plant's opening, but that they personally did not expect it to be a long-term success. One person said: "I hated to see Range Fuels fold, but it didn't surprise me."

While some people recognized that as companies employ new and developing technologies, there are bound to be failures, many interviewees discussed how they felt that Range Fuels was a scam from the beginning. As we discussed in the last chapter, people used phrases such as "we got took to the cleaners," "a lot of shady deals, people stuffing their own back pockets," and "shrouded in secrecy." People using these phrases felt that the closure of Range Fuels was not an accident or a simple technological failure, but rather the result of intentional deception by people who saw a way to illicitly profit from funds available for experimental renewable energy technologies. One community leader criticized both the heavy investment in experimental technologies and the aforementioned "dog and pony show" of their grand opening:

> Eighty million dollars and they don't know if it works. When you build a sidewalk, you should know what's on the other side . . . There was too much show and tell [with Range Fuels]. I had a gut feeling something was wrong at the grand opening of Range Fuels.

As noted, several people questioned the theatrics at the plant's grand opening, suspecting that the tanker that was supposed to be full of cellulosic ethanol was actually empty. Another person, a forester, said: "It was so staged—the truck was probably empty. Maybe it was a truck full of hops." Another forester said that he was gifted by company representatives a bottle of "ethanol" produced at the plant; he used air quotes to indicate his doubt about what he was actually given.

Several people specifically mentioned that many of the funds that Range Fuels received came from taxes, meaning that everyone in the country essentially had to pay, while some "very smart people" profited. This, understandably, led to a lot of anger directed toward this particular company, and by extension, to renewable energy in general. And again, much of this anger fell along party lines, since many of the politically conservative people with whom we spoke felt that progressive Democrats were allowing taxpayer money to be wasted on technologies that were still in the experimental phases. One community member said of Range Fuels:

> They got all this government money and then went belly up. When it sold, taxpayers knew their money was not too good . . . People had a bad taste in their mouths when it went bankrupt. It makes you think all they are after is my tax money.

Another claimed that Range Fuels had a lot of "trade secrets," citing a report from the newspaper in Macon, Georgia, which allegedly reported that "information was deleted from the Range Fuels report." We could not find this story in the newspaper, but the man continued to say: "Range Fuels was shrouded in secrecy . . . . This is tax money." There was so much anger in the community after the plant's closure that one person, a community leader, said: "Biofuels is a bad word around here now in Treutlen County. Everyone thinks we got took to the cleaners by Range Fuels."

In addition to the anger, blame, and sadness that people felt when the plant closed, they also mentioned the fears that people had when the facility was operational (and during the period between when it was announced and when it began production). A number of people with whom we spoke expressed concerns about ecological risks. Interviewees expressed concerns about potential negative ecological effects, including air pollution from the facility itself and increased traffic. Several people questioned how much water the facility would use and the temperature of the water coming out of the facility. Others worried about the environmental issues associated with using logging residues for biofuels, as these residues serve important ecosystem functions for wildlife and tree health.

Several people also mentioned health and safety risks associated with the Range Fuels plant. Some mentioned "rumors" that the facility was unsafe (specifically fears that it would "blow up"), and they talked about how these rumors were spread. A former Range Fuels executive told us these rumors were spread online, but we also found them circulating within the community. Multiple people mentioned the fact that a new school was being built close to the facility, and they expressed concerns about possible explosions, as well as increased pollution from the facility and about the dangers of increased logging truck traffic along the bus routes:

There have been studies done about the danger zones around a facility like this one, and if it blows up, the school is in a danger zone. The school is also in a dangerous location if one of the hauling trucks explodes on Hwy 29.

There needs to be serious investigation in how the school was built there, when there was so much opposition. There needs to be some head rolling for that.

People are worried, I'm worried, about gas leaks, explosions, truck accidents and more fumes from trucks. If there's an explosion, the wind could drift all kinds of stuff to the school. Others are saying this too, but now it's too late.

Many of these statements, and similar ones, were expressed by Black community members, though these concerns did cross racial lines, especially when we interviewed people who actually lived close to the facility.

Other people addressed concerns about the school but framed them dismissively as "rumors." For example, one man said: "There are rumors out there. One that caused an uproar was about the new school. The rumor was that it would blow up and kill children." Several other people also told us that some members of the community were just opposed to the expenditures required to build a new school that they did not feel was necessary, and that Range Fuels was essentially a scapegoat to try to stop or delay the school's construction. Several interviewees mentioned that Range Fuels provided evidence of the safety of the facility for the community, especially for the school, and that fears about health and safety were unfounded. Others mentioned that most of the opposition was coming from just a few people and that most people in the community were not concerned about the health and safety issues. Rather, some people claimed, they just did not want the new school.

There were plans for a new school to be built close to the plant. It was an excuse for the Board of Education not to build a new school. A bunch of NIMBYs[5] cited public health and safety issues—what if it exploded?

There were complaints and concerns about siting the new school near the Range Fuels facility: possible explosion, pollution, traffic, things falling off trucks . . . We've got some "engineers" here—they ain't got no more than graduated from Treutlen High, and now they's world-renowned engineers.

You got a bunch of nuts everywhere. But you could count them on one hand. Some people said it would blow up, wipe Treutlen off the map. Kill all the kids in school . . . Some people didn't want the new school. There were four or five people fussing.

Or, as one person said: "Some people used Range Fuels as an excuse not to build the school. Some are just against a plant in Treutlen County." Other people mentioned that some landowning families had profited by selling land to the county for the school and for the facility.

So it appeared to us that some people in the community were more opposed to the school than to the plant, while others were more opposed to the plant. Some, it also appeared, were opposed to both. In any case, it was clear to us that while many of the fears were real, and widely shared, there were also a lot of intracommunity dynamics involved that got framed as opposition to the bioenergy plant. The siting and (brief) operation of Range Fuels in Soperton also made a lot of these intracommunity dynamics clear by bringing a lot of tension between subpopulations in the town and county, and even between (and in some cases among) prominent families to the surface.

LanzaTech: New Company, New Opportunities, New Risks?

On January 3, 2012, the New Zealand-based company LanzaTech purchased the Range Fuels facility at auction for US$5.1 million and renamed it the Freedom Pines Biorefinery. LanzaTech retrofitted the facility for use as a research and development facility that focused mainly on chemicals produced using proprietary microbes and synthetic biology starting around 2013 (Schill 2014). Both companies have aimed to produce ethanol from syngas made from woody biomass, but LanzaTech was using a proprietary microbe in contrast to Range Fuels' use of catalysts (Bomgardner 2012). LanzaTech also had the intent to also pursue the production of jet fuel, which it is producing now, in a collaboration with British Airways (Burton 2021). However, at the time of the research, LanzaTech had recently purchased the old Range Fuels site, and company representatives were trying to differentiate their company from the previous owner.

Several interviewees commented on the differences between Range Fuels and LanzaTech in their intentions and preparedness as well as in the way they interacted with the community. Although many community members were even more skeptical about bioenergy development after the failure of Range Fuels, they still noted differences between the two companies and largely felt that LanzaTech had a better chance of success. One community member noted that

> I was not impressed with the first wave of biofuels people [Range Fuels]. They were money people. They move money around in a big way. The second wave [LanzaTech] is impressive. They are here to do something. They are interested in ongoing business.

Another person stated that because LanzaTech arrived with more of their own money (and less of the U.S. taxpayers'), they were more invested in their own success—which in turn proved to people in Soperton that they were more honest and better equipped to succeed:

> People say Range got it over on us and the federal government . . . People are leery. We'll see what happens. LanzaTech looks more organized. There's more money on the private side, not as much reliance on federal money.

Some people were still enthusiastic about bioenergy as a potential driver for economic growth in Soperton and optimistic about the future of the area in general. It should be noted that community leaders felt more optimistic than anyone else we interviewed. One said: "My gut feeling is optimistic . . . People are starting to buy homes in the county again. I don't want the town to be a ghost town . . . This could help the community." Another noted that "hope slipped away" when Range Fuels failed, but that "when failure happens, someone has to pick it up. We just welcomed LanzaTech to Treutlen County. They will pick up in some areas where Range Fuels left off."

However, some people were not at all confident that LanzaTech would be any more successful than Range Fuels was, and they expressed ongoing skepticism about wood-based bioenergy in general. A community member said: "Large landowners were excited . . . It's a different story now. I'm still skeptical. There's more distrust of bioenergy now because Range Fuels didn't follow through." Also expressing skepticism, a forester told us: "We hear a lot of rumors. I'll believe it when I see them buying wood and fuel coming out the other end." A farmer told us he was not just skeptical, but outright pessimistic: "Since the sellout and buy-back, people are pessimistic . . . It [LanzaTech] would be a boon. I hope it works. I'm all for it. But I'm pessimistic." Some were wary about this company in particular, namely because of the involvement of the same prominent cleantech investor in both Range Fuels and in LanzaTech (Bomgardner 2012). A forest landowner said:

> I'm not banking on it [LanzaTech] . . . The same guy that owned it before sold it to himself. He didn't lose money on it—the investors did, and we the taxpayers did. He's probably going to get the rest of those government loans and then sell it in pieces of stainless steel to the junkyard.

This statement echoes the rhetoric about Range Fuels being "snake oil," only with one person profiting both coming and going.

We interviewed several people who work with LanzaTech in different capacities (design, procurement, administration), and they all commented on how they have learned from some of the mistakes made by Range Fuels and were actively trying to do things differently, both technologically and socially. They all expressed satisfaction that they were proving success in areas where Range Fuels failed. One employee told us: "People here rabidly support what we're doing. They [Range Fuels] didn't deserve the media it got . . . Range Fuels only had three months of startup, and no money. We did well for what we were faced with." LanzaTech did three main things differently within the community in response to mistakes made by Range Fuels. First, it established a community advisory board. Range Fuels also had one, but it was widely criticized as being ineffective, most of all by the people serving on it. An employee of LanzaTech told us:

We live with the legacy of Range Fuels . . . We've heard people say that Range Fuels was secretive. We learn from that. We've set up a Freedom Pines Advisory Board to communicate with the public.

One of the main points the company emphasized, and spread through the members of the advisory board, was that LanzaTech uses a different technology than Range Fuels did and that poses less risk of "blowing up." They made sure to reassure community members that

We have all the safety protocols. The school is outside of our exposure area . . . LanzaTech uses much less pressure than Range Fuels did. Range Fuels ran pressure at 1800–2000 PSI. LanzaTech runs at about 40 PSI.

They understood the need to address people's fears directly and explain why their technology was different, and safer, than that used by Range Fuels.

Second, LanzaTech noted the criticisms that Range Fuels did not hire local people and made a commitment to do so. Many interviewees seemed to believe that they would follow through on this promise. A forester connected with the facility said that "Hopefully, LanzaTech will employ sixty plus people. They want to hire locals if they can. Range Fuels subcontracted everything, even security, cleaning. LanzaTech will hire locals." Other community leaders also believed that LanzaTech would draw much more heavily from the local labor pool; one stated: "It [LanzaTech] will employ sixty people at first, up to two hundred—that's a lot for a small place like Treutlen County. They promise to hire mostly locals."

Third, LanzaTech did not negotiate with the state, county, or city for economic incentives to locate in Treutlen County. It did not need to, as it had private funding, but this was perceived by many as another distinguishing factor between the two companies. People were especially upset about the tax abatements that Range Fuels had received, and they looked at LanzaTech more favorably for not pursuing these. Two different community leaders mentioned this explicitly:

Range Fuels got a twenty-year tax break, which was uncalled for . . . LanzaTech doesn't need a tax break because they bought it for $5.1 million . . . LanzaTech didn't get any tax breaks. That wouldn't be fair to taxpayers, to American citizens. LanzaTech didn't ask for a tax break.

People here have a better outlook now than before . . . The county gave a twenty-year tax abatement to get them [Range Fuels] here. LanzaTech invested out of pocket, didn't ask for an abatement. That would eliminate 90 % of their liability. They'd have no risk . . . LanzaTech may only have $5 million invested. Range Fuels already spent $60 million. How would you feel about them spending taxpayer versus stockholder money? They're in much better shape publicity-wise than Range Fuels.

The perception among a number of community leaders and community members was that a large, privately funded operation such as LanzaTech could in time bring the promised development to the economically depressed and forest-dependent town of Soperton, or at least not contribute to further decline.

Even the Dairy Queen Shut Down

Almost every town in the southeastern United States has a Dairy Queen, and the Dairy Queen is emblematic of small-town life. Since its first restaurant opening in 1940, its reputation as a hotspot for family outings, lunch meetings, and teen date nights makes it a social fixture and an important cultural institution. In Soperton, the closure of their Dairy Queen in 2011 represented not just the declining economy of the town but also the cultural losses associated with outmigration of young people to more economically prosperous towns and cities and the death of traditional industries and ways of life. Almost every person we interviewed in Soperton mentioned the loss of the Dairy Queen:

Recently, we even lost our Dairy Queen.

It was devastating when the Dairy Queen went away.

I visualize the way things were, what used to be here, even twelve years ago. We had a Dairy Queen. There's no more Dairy Queen here.

Soperton could be a center hub for several counties . . . I can see a Walmart here. There's no fast food. The Dairy Queen picked up and left. Downtown is majorly dilapidated. The roofs are falling in. But it's all run on a buddy system.

Treutlen County was all turpentine and farm. The community can't function now. I guess all small communities have a hard time. Here there's a lot of unemployment and drugs. Shops come and go like birthdays . . . It's bad here. Our Dairy Queen closed.

It's very sad, every southern town needs a Dairy Queen.

It was real sad when the Dairy Queen shut down.

The Dairy Queen in Soperton shut down during the year that Range Fuels was briefly in operation and then closed. Although the failure of Range Fuels and the closure of the Dairy Queen were not directly related, the two events are tied together in the minds of many community members: if Range Fuels had brought the development that it had promised, the Dairy Queen would be thriving, as would other establishments. Some people did link the two events, and one commenter in an online newspaper in Florida said, "Range Fuels brought so little of the promise to the table the Dairy Queen closed" (Ryan 2012).

When the Dairy Queen shut down and was not able to reopen, it became a symbol of disillusionment, iconic of the lack of the ability of the community to rebound from both economic and social and demographic losses. The risk is not just losing the economic base to support a fast-food restaurant; the loss of an iconic cultural institution also symbolizes a loss of resilience and breeds nostalgia for a past that was more vibrant and promising. Some people did express hope that one day industry would come to Soperton again and that a company like LanzaTech could help to revitalize this small-town economy by hiring local people. One man, an educator, said:

> They [LanzaTech] need to hire some people from Treutlen County. We have some good people here. They hired a few, but most were brought in from other places and they moved to Dublin, not Soperton. I don't blame them. We don't even have a McDonald's here . . . We don't even have a Dairy Queen anymore.

This quotation illustrates the aforementioned chicken and egg situation: Employment at new industries could bring back the Dairy Queen, but why would industry come to a place with no Dairy Queen? This is a conundrum felt deeply by many residents, and for many, the concurrent failures of Range Fuels and the Dairy Queen cemented the fate of Soperton as another southern town that is now, as one interviewee phrased it, "a cemetery" (see figure 3.2).

Our research shows that several issues related to bioenergy technology, wood and energy markets, energy policy, and social and economic impacts operate at multiple scales simultaneously, both literally and discursively. They converge on the ground in local places where bioenergy is sourced and produced, and we contend that renewable energy development can introduce new uncertainties and new risks into energy systems and local communities. These risks and opportunity costs to local communities are often overlooked in imaginaries promulgating bioenergy as a driver of economic and ecological resilience. Using the example of the Dairy Queen, which has cultural meaning to community members transcending its functional status as a fast-food restaurant, we show that the closure of this iconic establishment is both a symbol of the risks borne by local communities as new technologies develop

Figure 3.2   The Closed Dairy Queen in Soperton, Georgia, 2011. *Source*: Credit: Sarah Hitchner.

and also a visible indicator of economic decline, which many people perceive as an omen for further economic distress in the community.

Forest-dependent communities in the southeastern United States have long experienced both dramatic fluctuations in the wood product industries (in terms of supply, demand, and labor and mill accessibility) and failed promises to bring new economic development and employment opportunities to rural communities. As a result, many people in these communities are skeptical about advancements promising to provide jobs and new markets. However, bioenergy representatives and many local and national politicians are heavily promoting sociotechnical imaginaries in which bioenergy is a potential solution to rural poverty, dependence on foreign countries for fuels, and loss of markets for wood products. Prior to a bioenergy facility being built in a rural community, some community members feel that this is feasible, while others do not.

However, failing bioenergy projects can continue the cycle of disillusionment and distrust of the sociotechnical imaginaries pushed by government agencies and officials and industry leaders. Failures of specific bioenergy projects like Range Fuels have not only cost the states, counties, and towns in which these facilities were located millions of dollars in grants and loans; these failures have also led rural people already wary of government intervention in industry to feel even more betrayed by government policies that seem to benefit a few at the expense of the public in general and local communities in particular. These failures can crystallize opposition to bioenergy that has emerged out of these failures and a general distrust of government intervention in markets, which themselves are embedded in a political imaginary of ordinary people and communities suffering from collusion between government and business interests.

The fundamental disconnect between positive and negative views of bioenergy reveals the divergence of incommensurable imaginaries in which wood-based bioenergy is, or is not, a potential solution to current and future energy problems. It also reveals differences in personal experiences with bioenergy in local contexts and policy arenas; for example, people promoting bioenergy through industry development or policy creation often do not live in rural communities in which bioenergy facilities are located and thus do not feel the effects of these facilities' success or failure. In fact, they may never see them firsthand or even think about them. The imaginary that promotes bioenergy development as a driver of economic resilience for rural and impoverished towns also downplays the risks, including economic opportunity costs, that community members experience and perceive when expensive new facilities with experimental technologies are located in these areas. This case study demonstrates how, in a place where forestry is deeply embedded in community history and culture and where the success of a wood-based bioenergy facility was supposed to "put Soperton on the map," its failure instead exacerbated an ongoing economic and social decline that resulted in the loss even of the Dairy Queen.

Talking about Bioenergy: When Imaginaries Collide in Forest-Dependent Communities

As we identified the main themes that came up in our interviews with local community members and landowners about bioenergy in relation to their forests and communities and then analyzed these narratives, we realized the complex intermingling of the various elements of imaginaries. Part of the reason is because of the multi-scalar nature of bioenergy discussions, policymaking, and on-the-ground development in specific locations; what we saw in local communities did not always translate from the bioenergy imaginaries being promulgated from above. For example, the bioenergy plants in our study sites were designed to use woody biomass harvested from nearby forests and to produce products for national and international markets. Notably, all three plants were stand-alone plants that did little to take advantage of efficiencies and synergies with other wood product industries (see Aguilar 2014). Development of each of these plants was high profile and had stated goals of energy security, renewable energy, and rural development, thus fitting into a sociotechnical imaginary which represented a powerful vision of a new energy future strongly supported by the state that advocated for and supported new technological development that would entail significant changes in social and economic systems. The vision claimed to be able to address multiple issues, including energy security, rural development, and climate change through the use of the South's forests in ways that are sustainable and renewable. It minimized risks and uncertainties while emphasizing the unbounded potential to develop a new, bio-based industry. It involved strong state support through discourse, policies, and financing, but numerous non-state actors are also involved in developing, promoting, and supporting the vision. Driven by this imaginary, bioenergy facilities were proposed and developed in multiple communities in the southern United States. At the same time, the imaginary relied on a vision of the public good that was defined by narrow interest groups without widespread public engagement, leading to questions about its durability and motivating force (Kim 2015; Hurlbut 2015).

In accordance with recent literature on energy imaginaries, we recognized that this imaginary is not homogenous or universally endorsed (Delina and Janetos 2018; Eaton, Gasteyer, and Busch 2014; Burnham et al. 2017; Smith and Tidwell 2016). Counternarratives to contest it, which emphasized negative environmental impacts and wasteful government expenditures, have emerged at the national and regional levels. Following Strauss (2005, 2006, 2012), we endeavored to trace elements of the national bioenergy imaginary and related counternarratives in talk about bioenergy in communities in the southern United States where bioenergy production facilities have been implemented in response to the national bioenergy imaginary and related policies. Our findings focus on liquid fuel and wood pellet bioenergy plants developed at a time of high enthusiasm for bioenergy. By examining the ways that cultural models related to bioenergy imaginaries are repeated, contextualized, contested, and reshaped in the talk of community members with varying levels of engagement in bioenergy development, we have endeavored to understand how the national bioenergy imaginary operates in local social and cultural systems.

As we expected for dominant discourses and related development (see, for example, Schelhas and Pfeffer 2008), we found that the dominant national bioenergy imaginary played an important role in structuring the way local people talked about bioenergy. In particular, the themes of energy independence and security and rural development resonated at the local level. Rural southern U.S. communities tend to be politically conservative, and arguments about independence from Middle Eastern oil sources, prioritizing domestic resources and production, and supporting and protecting soldiers are very much in line with the way people already think. Support for domestic industry and shrinking employment in traditional wood product industries lead to support for bringing new industries to these communities. These elements of the bioenergy imaginary appear most strongly in the talk of local promoters of bioenergy, including people associated with related industries and local elites involved in economic development.

Within the broader population in our research communities, we found people looking more closely at the contribution of bioenergy plants to energy production, actual job creation, and improvement of wood markets. Many of these material benefits failed to materialize, seemed precarious, or went unnoticed in local people's experiences. This led to some skepticism about bioenergy development. This skepticism conformed in many ways with the larger conservative political discourse, prominent in the southern United States, that is suspicious of government spending and subsidies. Many people drew on this discourse in their belief that bioenergy development was a scam to benefit elites, a belief that was also fed by widespread disbelief in climate change. Others in the communities countered this perspective by pointing out the many ways that other energy industries and infrastructure were subsidized in the United States, arguing that this was more a leveling of the playing field. This incommensurable contrast in views, between seeing the bioenergy development in their communities as wasteful government spending to benefit elites and promoting development within existing structures and patterns, represented an important cleavage and tension within communities. The former group rejected many of the key components in the national bioenergy imaginary.

The communities we worked with can be characterized as forestry or wood products communities. All were or had been home to mills of one type or another and were home to many family forest owners. We know from other research that Southern family forest owners value forests broadly (Butler et al. 2016). Even when they manage tree plantations, forests are at least as important to them for wildlife and hunting, aesthetics, and family heritage as they are for economic return from timber. People tended to view wood-based bioenergy development from a perspective rooted in their community's recent experience with forestry and forests. They were accustomed to timber harvesting, transport, and processing, and there was little evidence that they were influenced by environmentalists' messages about bioenergy being a threat to forests. At the same time, they had a relationship to forests that was much broader than as a source of timber. There was a moral dimension to the ways they described the forests owned by their families and surrounding their communities. The presence of forests contrasted their communities with urban areas and was linked to religion and healthy lifestyles. While promoters of bioenergy saw forests as raw materials and talked of planting new biomass feedstock and intensifying production, forest owners, family owners in particular (but industrial and corporate owners as well), were generally satisfied with the existing native pine plantations. Through this lens, they saw their forest management as renewable and sustainable, and showed very little interest in meeting other definitions of these terms related to national and international markets as would be required by bioenergy development.

What emerged from our research is a largely coherent and shared story about recent bioenergy development. Our respondents generally valued the communities they live in, although there was some disenfranchisement along racial and class lines. They recognized that they are forest product communities surrounded by forests and tree plantations, and that their most likely paths for economic development would involve better wood markets and jobs in wood-processing facilities. Bioenergy development was generally seen as a good fit for their communities, and they were receptive to arguments for energy independence and security. At the same time, they drew on other discourses in their suspicion of climate change, government subsidies, and policy interventions in a market economy. These other discourses combined with uncertainty around pellet markets and the failure to develop economically viable liquid fuels to feed skepticism about local bioenergy development. Simultaneously, environmentalist counternarratives stating that bioenergy would lead to forest loss and degradation had little influence. Community members engaged in timber harvesting and the forestry industry in ways that suited their aesthetics, lifestyle, and household economies shared broad support for forestry but had little interest new silvicultural systems or meeting outside standards of forest management. Primarily, local communities wanted jobs and improved markets for wood from their conventionally managed pine plantations.

In focusing on the salience and nature of the national bioenergy imaginary in local communities, our research draws attention to the scalar and translational dynamics that characterize bioenergy development in the U.S. South. We found acceptance of the call for energy security and independence, rejection of the climate change justification, and ambivalence about government promotion and subsidies. The imaginary became diluted at the local level in that it addressed fewer broad goals, and it was in some sense co-opted as it was exploited primarily for local development and wood markets. Our findings have some similarities and some differences from what other researchers have found. We did not find differing niche bioenergy imaginaries emerging (Burnham et al. 2017), nor did we find the different interpretations of "wood for energy." The bioenergy developments in the communities we studied came from the outside and were seen as vehicles for economic development compatible with existing forestry communities and forest management. This likely reflects differences in the nature of bioenergy facilities and the role of timber harvesting between regions in the United States. Powerful interests in communities were associated with bioenergy development and emphasized job creation and minimized risks; however, we found more people willing to speak out against this. This was probably due to the nature and uncertainty of the facilities in our study communities, as well as perhaps a greater willingness for anti-government activists to speak out in a time of anti-government sentiment. Our ethnographic methods may have also enabled us to reach more people who were out of their community's mainstream. Finally, we did not find links to environmentalist counternarratives and imaginaries or other social movements (Kim 2015), instead finding limited local interest in changing forest management to meet the demands of bioenergy.

Many different actors, some local, drew on the national bioenergy imaginary, both because they saw in it a solution to multiple problems and because their own interests fit within it. Bioenergy development was aggressively promoted by a variety of interests, and communities desperate for development were asked to commit to bioenergy plants based on uncertain technology and policies that could change. Very optimistic scenarios and inaccurate descriptions of raw materials were provided to local communities. Local people brought their own interpretations to bear on the projects, which diminished the resonance of the national bioenergy imaginary. Non-elite locals had few opportunities to be heard during bioenergy development, and their views were not incorporated into visions or development. These issues were exacerbated by the particular nature of bioenergy development in the communities in which we worked, which was implemented through outside government and business initiatives. The limited success of the liquid fuels plants was also in part due to generally unfavorable policy and market environments for renewable energy. The expectation of a mechanism to regulate carbon emissions and address climate change, combined with relatively low natural gas and gasoline prices, clearly shifted during our research. Nevertheless, climate change and geopolitical turmoil will not go away, and a resurgence of interest in renewable energy and bioenergy, including cellulosic biofuels, seems likely at some time in the future (Lane 2017). Bioenergy imaginaries can be powerful cultural resources to motivate action, but our research shows that top-down imaginaries can become diluted and lack motivating power among key participants in bioenergy development at the local level. Failure to account for local interests and values may limit the effectiveness of imaginaries in promoting bioenergy development.

Imagined energy futures have technical, social, and moral elements and are rarely universally shared, and their implementation plays out in the context of a variety of other factors including national energy security, political economy, and environmental justice. Within individuals, communities, and societies, energy visions and their implementation can fragment across value spheres and require trade-offs or even outright acknowledgement of incommensurability, meaning that energy transitions are complex processes (sometimes called wicked problems) that require continual attention to address interpretations and tensions. Our analysis of bioenergy development in the southern United States, and the various ways that people talk about it, highlights some of the issues that can develop when diversity and social complexity are not taken into account when envisioning and implementing an energy imaginary, emphasizing the importance of both social research and inclusive approaches.

As noted by Molyneaux et al. (2016), broad-scale economic resilience is dependent on resilience of energy systems, and analyses of the resilience of these systems often overlook the role that renewable energy technologies can play in building adaptive capacity to uncertainties, in this case, fluctuations in oil prices, shifts and relocations of wood products industries and markets, and advancements in bioenergy technologies. Our research shows that several issues related to bioenergy technology, wood and energy markets, energy policy, and social and economic impacts are operating at multiple scales simultaneously, both literally and discursively, and have direct impacts on local communities experiencing bioenergy development firsthand. We contend that the imaginaries emerging from actors and systems that operate on large scales create effects in local communities that may not be very resilient. These imaginaries are sold as mechanisms to increase community resiliency, but they may in fact erode it further by appropriating resources such as tax abatements and limited industrial sites, which in turn undermines a community's ability to pursue other economic opportunities to strengthen and diversify its economy.

THE FOREST IS IN YOUR BLOOD

We find it necessary to emphasize the importance of the ecological context in which these bioenergy imaginaries play out, and, in particular, people's emotional and cultural connections to it. The relationship between people and the forests of the southeastern United States has been complex and dynamic. The landscape is a historical palimpsest that begins with Native American occupation, resource use, expulsion, and genocide and then moves toward exploitation of both natural resources and slave labor by White colonists. Later, much of the forested region originally dominated by longleaf pine was cleared for agricultural plantations and large-scale monocultures. Of that cleared agricultural land, much has since reverted to forestland, both through natural regeneration and intensive silviculture of pine plantations. These changes in the forests of the region both affect and are affected by changes in the different valuations placed on trees and forests by different people.

These changes in ecosystems and in human values related to forests have important implications for forest management strategies employed in particular by private landowners and industrial timber plantations, as so much land in the South is privately owned and managed. When new forest uses, such as wood-based bioenergy, make the transition from a possibility to an actuality, people must analyze and, in some cases, reconfigure their management strategies and goals. As we have seen, some landowners are more willing to make changes in terms of species or rotation than others. We've also seen how the success or failure of wood-based bioenergy plants, or the perception of future success, is an important indicator of people's willingness to embrace new uses for southern trees. Attention to the way that people talk about trees and forests can give important insights into how people feel that bioenergy will fit into the socio-ecological landscape of southern pine forests. As a farmer and forest landowner in Columbus, Mississippi, told us: "Timberland is something you need. I like to see good timber. It's pretty . . . . The forest is in your blood."

NOTES

1. The North Carolina State University School of Forestry describes "site index" in this way in its "Woodland Owner Notes" website: "The collective influence of soil factors will determine the site index for a particular tree species on a given soil area. Site index is the total height to which dominant trees of a given species will grow on a given site at some index age, usually 25 or 50 years in the Southeast. Dominant trees are the tallest trees in the stand. If it is stated that an area has a site index for loblolly pine of 70 feet at 50 years, then we expect loblolly seedlings planted on that area today to be 70 feet tall in 50 years. Index age and tree species must be stated when referring to site index because the site index of one species will be different from the site index of another species growing on the same area. There IS a close relationship between site index and timber yield. Volumes of merchantable wood increase with improvement in site index" (https://content.ces.ncsu.edu/forest-soils-and-site-index).

2. Conservation Reserve Program: "CRP is a land conservation program administered by FSA [Farm Service Agency]. In exchange for a yearly rental payment, farmers enrolled in the program agree to remove environmentally sensitive land from agricultural production and plant species that will improve environmental health and quality. Contracts for land enrolled in CRP are 10–15 years in length. The long-term goal of the program is to re-establish valuable land cover to help improve water quality, prevent soil erosion, and reduce loss of wildlife habitat" (https://www.fsa.usda.gov/programs-and-services/conservation-programs/conservation-reserve-program/).

While many landowners have profited from the CRP programs, and many foresters encourage it, some oppose it because it limits the raking of pinestraw for the first ten years after a longleaf tree is planted. One forester and landowner told us that this rule was proof that "The government is more interested in wildlife and soil and water than the timber business. It don't make sense, but you know how the government is."

3. Diameter at Breast Height: common unit of measurement for tree size, roughly 4.5 feet from the ground.

4. Southern yellow pine (SYP) consists of a group of pine species that are especially common in large-scale timber production across the southern region. The four main species considered SYP are longleaf pine, shortleaf pine, slash pine, and loblolly pine.

5. NIMBY means "not in my backyard," and some people refer to people who protest new developments on this basis "NIMBYs."

Chapter 4

What's Climate Change
Got to Do with It?

The Relevance (or Not) of Climate
Change to Perceptions of Bioenergy

CLIMATE CHANGE: LIVING IN THE INTERSECTION
OF SCIENCE, CRISIS, AND CRITIQUE

At a bioenergy conference in Georgia in 2013, a keynote speaker from Denmark emphatically stated that "Climate change is happening whether we believe it or not." She went on to say that transitioning to a no-net-carbon economy is a relatively inexpensive process; once subsidies for fossil fuels are removed, renewable energy sources will be competitive in the global market. Woody biomass from the southeastern United States, she said, will be key to reducing global greenhouse gas emissions and lightening humanity's carbon footprint. At another bioenergy conference in Mississippi in 2013, a bioenergy industry representative began his remarks by saying, "Half the people in this room believe in climate change, and half don't. I'm not going to get into that conversation. Either way, reducing CO_2 is a good thing because it uses less energy." These remarks demonstrate the contrast between many European nations (and most other nations around the world), which take the irrefutable fact of climate change as a starting point for discussions about what to do about it, and the United States, in which climate change has been highly politicized and whose existence is treated as a political hot potato instead of a matter of scientific consensus. Recognition of this disjuncture was expressed by a speaker at a bioenergy conference in Georgia in 2012, a procurement forester for a pellet facility in Georgia whose products supply European markets. As he put it, "Telling a European that climate change is a hoax is like telling a Southern Christian there is no Jesus."

These remarks reveal the complicated relationship between climate change and bioenergy in the U.S. South. Mitigation of climate change is one element of the pro-bioenergy imaginary, along with other complementary societal benefits such as national energy independence, rural development, poverty alleviation, and expanded wood markets for landowners. At the national level, reliance on renewable energy to address climate change has been a primary justification for public investments in bioenergy. However, climate change skeptics abound in the conservative rural South, and speakers at regional bioenergy conferences must often try to make a broad appeal to the audience by emphasizing other important benefits of bioenergy beyond climate change mitigation. Here we use examples from our field sites, both in the communities and at regional bioenergy conferences, to analyze the relationship of bioenergy to climate change across scales, focusing on what happens to the bioenergy imaginary at the regional and local levels when it encounters climate change denial. We look at how the bioenergy imaginary does—or does not—accommodate an absence of one of its key elements. Just as important is the question of how belief or disbelief in climate change affects views about bioenergy and specific components of it such as sustainability certification, national energy policies, and bioenergy regulatory processes. Opinions about these issues, like most all other issues in the South, are steeped in political allegiances and racial dynamics.

Without needing to invoke the cliché image of the starving polar bear on the isolated fragment of floating ice, we can acknowledge that the planet is in a precarious state. The science is clear and unambiguous, and the scientific consensus is strong that (1) the climate is indeed changing, (2) this is not a good thing, (3) human activity is the main cause, (4) humans need to do things differently in order to avert ecological disaster, and (5) the effects of climate change are neither consistent nor universal: different biomes and human populations will be affected by climate change in different ways and to different degrees. However, discussions about climate change are anything but straightforward; they are instead situated at the muddy intersections of science, the crisis narrative, and post-truth partisan politics. First, we must make sense of the mountains of data about historical changes in climate patterns, carbon cycles, and sea levels, among other factors. What is the science telling us about what is happening? What is the level of agreement among scientists and data sets, and how are future projections about climate change made and shared? Second, we need to accurately assess the severity of the crisis. How much trouble are we in, and what should we do about it? Crisis narratives have driven environmental movements regarding other ecological crises such as deforestation, acid rain, ozone holes, and biodiversity loss, and these narratives have been promulgated by a number of actors with a variety of motives and varying levels of understanding of the science behind them. Third, it is commonplace in environmental discourse, especially when crises are involved, to invoke the "we"; that is, "we" need to do something drastic to avert imminent disaster. The questions of interrogating the "we" and deciding who "we" involves are central in determining the fate of the planet. Which "we" is entitled to do something about climate change? Which "we" has the privilege to be able to ignore it (for now), and which "we" is most affected by climate policies, or lack thereof? These are all consequential questions and should be at the very center of conversations about not only the scientific evidence regarding the causes and effects of climate change, but also the social reasons for and consequences of climate change denial and differential vulnerability to climate changes.

Though there is overwhelming scientific evidence supporting the notion of anthropogenic climate change, it is not an irrefutable fact among many Americans, particularly in the rural southeastern United States. Climate debates are deeply embedded within a matrix of other discussions about power, race, and politics, and in this region climate change often serves as a symbol of polarization and partisanship. Though climate change is a major driver of bioenergy policy both nationally and internationally, it does not necessarily smoothly guide the development of economic and environmental policies, especially on regional and local scales. Rather, as science and technology studies (STS) remind us, science and governance are continually coproduced (Scott et al. 2014), and friction (Tsing 2005) occurs when different, possibly incommensurable, imaginaries collide, especially when the driving force of those is under debate.

But even when different imaginaries share the common premise that climate change is real and an important issue to address, they can differ in their view on whether bioenergy should play a prominent role in mitigating it. For example, the general pro-bioenergy imaginary that we have referenced throughout the book proposes that wood-based bioenergy in the South can simultaneously address climate change, rural development, energy independence, and ecological sustainability. Proponents of this imaginary claim that production and use of wood-based bioenergy products lead to increased forest cover and reduced emissions and pollutants compared to fossil fuels, and that the U.S. South, as a known woodbasket, has plenty of resources to fuel the nation (and other nations with fewer resources). In this imaginary, bioenergy is an important mechanism for mitigating climate change and a host of other benefits. In the other, bioenergy is an important mechanism for a host of benefits other than climate change. Either way, bioenergy is regarded as a good thing.

Both of these imaginaries are in contrast to an imaginary in which bioenergy is an obstacle to the main goal of climate change mitigation. This imaginary shares the same end goals as the main bioenergy imaginary; however, it includes climate change mitigation, but rejects the idea that wood-based bioenergy is the appropriate vehicle to accomplish them. Actors promoting this alternative version of the bioenergy imaginary, often representing or influenced by nongovernmental environmental organizations, claim that wood-based bioenergy can actually exacerbate climate change when calculating the full life cycle of bioenergy products (i.e., production, transportation, and consumption). They are particularly vocal against the use of domestic forest resources for foreign markets, both because transoceanic transport negates any climate benefits and because the forested ecosystem here in the United States serves as a global carbon sink. These are not simply translation issues; they are distinct imaginaries.

Different actors—bioenergy and forestry industry representatives, environmental NGOs, scientists, landowners, community leaders, and community members—circulate sometimes conflicting or incommensurable ideas about the role that wood-based bioenergy can play in either mitigating or exacerbating climate change, and these actors do not always fit squarely within the confines of one imaginary. We find that the boundaries between imaginaries are porous at times, as ideas are passed back and forth between actors (as we explored in chapter 2). Bioenergy, one expert in biomass told us, "makes for strange bedfellows."

In addition, audience matters; what a bioenergy industry representative might say to promote her company at a regional rural development forum focused on local job opportunities would likely differ from what she might say when lobbying at a national level to increase mandated ethanol blending requirements; if talking to a group of environmental activists, the message would be tailored to focus on evidence of sustainability. As we have already discussed, speakers use words and phrases strategically in order to persuade, and the most persuasive speakers take care to appeal to the values that are most important to their audience.

When the values of some groups are not adequately addressed, it can render these groups powerless to shape the trajectories of the dominant imaginary (or imaginaries). For example, minorities and politically progressive community members are especially concerned about potential ecological repercussions of large-scale wood-based bioenergy development and about the environmental justice implications of climate change and unequal levels of social vulnerability to its impacts. However, what we found was that bioenergy industry representatives and proponents dismiss these concerns by saying that bioenergy development will happen "whether you believe in climate change or not." This framing sidesteps the debates and concerns about the issue of climate change and essentially renders them irrelevant. It also dismisses the concerns about the unequal distribution of costs and threats of bioenergy development on poor and minority communities, which is touted as a way to reduce harm from climate change. In other words, communities of color often bear the brunt of both the disease and its potential cure. Regardless, in places in the

U.S. South where wood-based bioenergy is developing, it is happening relatively independently of concerns about climate change. If the climate piece is removed, wood-based bioenergy may just become another wood product market; this would suit many rural forest landowners in the region just fine.

A BRIEF INTRODUCTION TO CLIMATE CHANGE SCIENCE AND POLICY

The Science of Climate Change

The evidence supporting the fact of climate change is irrefutable, based on innumerable, scientifically rigorous, peer-reviewed studies and reports, and particularly the assessments produced every five to six years by the IPCC (Intergovernmental Panel on Climate Change). While there is ongoing debate in the public sphere about the degree to which these changes are the result of human activity or natural variations—in the United States largely configured along partisan lines—there is broad consensus in the scientific community on the anthropogenic nature of climate change. We will not delve deeply here into the science of climate change, but to briefly summarize, greenhouse gases are gases (mainly carbon dioxide, methane, nitrous oxide, and fluorinated gases such as hydrofluorocarbons and perfluorocarbons) that absorb long-wave infrared radiation, preventing solar energy from escaping back into space. Through a series of ecological feedback loops connecting the earth and its atmosphere, these trapped gases have "thickened the earth's blanket" and raised the mean temperature of the Earth's atmosphere, causing the balance of the earth's atmosphere to shift in ways that threaten life on earth.

These greenhouse gases, each with its own Global Warming Potential (GWP) rating, are all rising exponentially, especially CO_2. There is solid scientific evidence showing that atmospheric CO_2 levels are currently well above historical levels, based on recent data taken from direct measurements and historical data taken from ice cores in Antarctica holding ice bubbles up to two million years old (Yan et al. 2019). The levels of CO_2 in particular have accelerated at an alarming rate since the Industrial Revolution; in 2019, the global average amount of CO_2 hit a new record high (409.8 parts per million), higher than at any point in the last 800,000 years (Lindsey 2020). A number of deleterious effects of climate change have been, and are continuing to be, recorded and monitored; the effects are both dire and not evenly distributed. Global mean temperature has been consistently rising since the 1880s, though some areas have gotten cooler. There is also widespread evidence that climate change is changing the intensity of extreme events: hurricanes are getting stronger, droughts are getting longer, and wildfires are becoming more frequent and intense. What is clear is that the climate is changing at different rates in different places.

While there are natural fluctuations in CO_2 levels, these mostly have to do with small variations in the earth's orbit around the sun. These fluctuations are normal, known to science, and accounted for in scientific modeling regarding historical changes in climate change indicators. What we are experiencing now is that "the annual rate of increase in atmospheric carbon dioxide over the past 60 years is about 100 times faster than previous natural increases, such as those that occurred at the end of the last ice age 11,000–17,000 years ago" (Lindsey 2020), and human activity is without a doubt the main culprit. Ice cores and other historical data show that changes in the global carbon cycle are caused by humans through activities such as deforestation and forest burning, industrial manufacturing, and the production and use of fossil fuels. It is also important to take into account the fact that methane leaks from natural gas extraction are major contributors to climate change, actually negating the majority, if not all, of the climate benefits of switching to natural gas from coal (Storrow 2020).

On "Uncertainty" in Climate Science

Key to understanding contemporary climate science is the concept of uncertainty. What bears noting is that when climate scientists talk about uncertainty, it has a different meaning than that circulating in public discourse. In normal conversation, uncertainty means, "I don't know," or even, "It can't be known." In scientific terms, the word points to a range of possibilities, the exact one of which is unknown. When extrapolating the range of possible outcomes related to climate change, climate scientists provide scenarios, based on models using data that suggest where the curves might be going if different variables come into play (Heal and Kriström 2002; Reilly et al. 2001, van Vuuren 2020). Depending on what happens, including feedbacks of which we may not yet be aware, outcomes may differ. Scientists are, therefore, uncertain about the final state. For example, projected sea level rise is no easy calculation. Sea ice is disappearing in cold regions around the world, and the Greenland ice sheet in particular is melting from underneath (Oswald and Gogineni 2012). The melting of these ice sheets, as well as the melting of permafrost in boreal regions, releases massive amounts of methane into the atmosphere, which further speeds climate change (Box et al. 2019; Guo et al. 2020). This melting, and subsequent methane emissions, combined with thermal expansion of ocean water, has cascading effects that can affect model outcomes. Due to still incomplete knowledge about hydrothermal cycles and feedback effects, there is resultant uncertainty about final outcomes, leading to various scenarios of the projected level of sea level rise by the year 2100. A conservative estimate might predict a rise of only five to seven inches/12.7–17.8 centimeters, while another may predict a rise as high as 8.2 feet/2.5 meters (Prandi et al. 2021; Sweet et al. 2017). There is no uncertainty that sea levels are rising; that is basic physics. But scientists are not sure how much it will rise over time. There are, again, a range of possibilities. When we take these scenarios of mean temperature rise and extrapolate these data to predict how they might affect sea level rise and inundation, and then overlay these scenarios onto maps of actual places, we can anticipate that coastlines around the world are in trouble. Climate change has the potential to raise sea level enough to cover a number of the world's most populous cities, which are also home to many of the world's most vulnerable people (Sweet et al. 2014).

While there are a number of different projections, reflecting the variations and uncertainties about the nature of the curve, the science agrees that climate change is real, and that it is very dangerous. Climatologist Marshall Shepherd compares uncertainty in climate science to the hurricane cones we see on the news projecting the possible path of hurricanes; they cover a wide area, and forecasters cannot say exactly where the hurricane is going to go. These "cones of uncertainty" are often misunderstood by the public (Broad et al. 2007; Shepherd 2017), though they are useful visualizations of the fact that the hurricane will go somewhere, and uncertainty about its path does not translate into uncertainty about its existence.

One question that has animated the climate change conversation in the public domain is the extent to which we have any certainty that what we are observing is anthropogenic. As noted, natural fluctuations in levels of carbon dioxide and other gases are known and accounted for in climate models, and the time scales on which carbon dioxide levels have increased since the Industrial Revolution far outpace previous cyclical increases. Shepherd (2011, 7) writes:

> The carbon cycle is a delicate part of the Earth system, and increasingly the human or anthropogenic footprint is apparent. In the contemporary discussion of climate change, one fact is very clear and not debatable; human activities have altered the carbon cycle, particularly the atmospheric component.

When we examine the primary drivers of climate change, it is important to emphasize that if we deny the anthropogenic causes of climate change, what is required is the proposal of a scientifically valid alternative explanation. In science, nothing can be proven, only falsified, and an alternative explanation to anthropogenic climate change requires a contrary hypothesis and valid and rigorous scientific data to back it up. No such alternative exists, at least none with any credibility.

How Climate Policy Gets Made

The general public's view of climate change and climate change policies is heavily influenced by popular media (i.e., books, films, and social media), public figures (i.e., musicians, actors, and politicians), political pundits (liberal, moderate, and conservative), and the opinions of people they trust. More highly educated people tend to have a more comprehensive and nuanced understanding of what climate change is, what causes it, and the range of its possible effects. However, what regular people, who are not policy-makers, often do not fully understand about climate change policy is "how the sausage is made." These processes are largely hidden from view, as discussions about climate change policy happen within the confines of distant institutions and international meetings to which most people are not privy. To gain a deeper understanding of how decisions are made, it is vital to understand how climate policy is constructed and enacted, and how those processes intersect with climate science.

Climate change is simultaneously a scientific issue, in that it requires rigor in terms of observation, analysis, and modeling, and a social issue, in that its foundation is the fact that it presents us with an existential threat to humanity and all other life on this planet. The whole edifice of climate science is premised around the idea of the science-policy interface (Lacey et al. 2018; Wesselink et al. 2013). There exists a binary in which scientists do science, not determine what policy-makers do and people do. Those decisions are relegated to elected officials and delegated authorities who set policy; that, it is understood, is where the domain of decision-making should be contained. However, these lines do blur, and some climate scientists have been important conduits of scientific information to the public and to decision-makers.

The key convening international body on climate change is the IPCC. It was formed under the United Nations in 1988 and "provides a framework for governments, scientists and IPCC staff to work together to deliver the world's most authoritative scientific assessments on climate change."[1] Figure 4.1 shows the organizational structure of the IPCC; the work of governance is overseen by the Secretariat, while the real work of the organization is done by the Working Groups and Task Force. These groups convene top scientists from all over the world who are doing the research; guided by an established set of principles and procedures, these researchers contribute their expertise on a voluntary basis. The three Working Groups and one Task Force as shown in the graphic are supported by Technical Support Units (TSUs). The Working Groups meet in a Plenary session that includes government representatives, while the Task Force is charged with calculating increases and decreases of greenhouse gases in the atmosphere. Together, these groups create Assessment Reports every six years or so, which present comprehensive

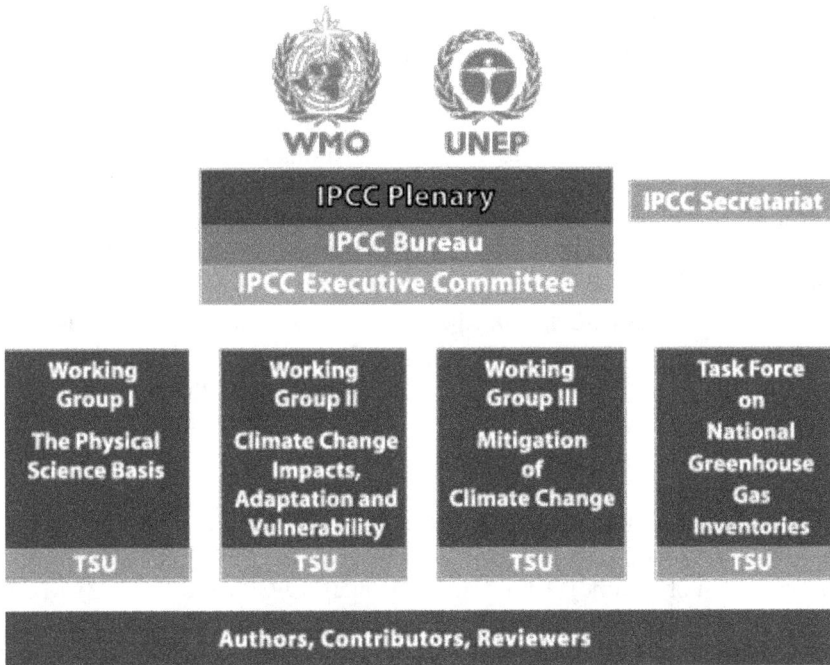

Figure 4.1 "How does the IPCC work?" from IPCC Website. *Source*: (https://archive
.ipcc.ch/organization/organization_structure.shtml).

information "about the state of scientific, technical and socio-economic knowledge on climate change, its impacts and future risks, and options for reducing the rate at which climate change is taking place."[2] They then release those reports, which become a matter of public record (Reilly et al. 2001).

The reports take stock of where we are ("observed changes"), outline possible future risks and impacts, and suggest different "pathways" for adaptation and mitigation. They also convey, using the best and most comprehensive scientific data available, a measure of the certainty of human causation of climate change, with increments of certainty calculated in percentages pegged to "certainty-calibrated qualifiers of confidence and likelihood" (Herrando-Pérez et al. 2019) such as "extremely likely" and "virtually certain" (Lewis and Gallant 2013). As each assessment has appeared since the first one was released in 1990, the level of statistical confidence in the anthropogenic cause of climate change has risen steadily. The first assessment did not quantify the likelihood of anthropogenic climate change. By the second assessment in 1995, the IPCC stated that human impact on climate was "discernible." The likelihood that anthropogenic climate change accounted for more than half of the increase in Earth's temperature since 1951 was considered "likely" (67–97% degree of certainty) in the third report published in 2001, and by

2007, the fourth report claimed it was "very likely" (at least 90% certainty). Another assessment was released in 2014, and it concluded that the degree of certainty of anthropogenic climate change was "extremely likely" (95%). In the Sixth Assessment, released on August 9, 2021, the degree of certainty approaches 100%, as it states that "human influence on the atmosphere, ocean, and land" is "unequivocal."

The year 1992 is a good starting point for discussions about the intersection between conservation and climate change. The United Nations Conference on Environment and Development (UNCED), also known as the Earth Summit or Rio Summit, was held then in Rio de Janeiro. One of us, Pete Brosius, was in the United States during the first half of the summit and conducting field-work in Malaysia during the second half, witnessing firsthand the different conversations taking place in the Global North and the Global South. At that time, Prime Minister Mahathir Mohamad of Malaysia was positioning himself as a spokesperson for Global South. The Rio Summit brought together world leaders to discuss, among other topics, climate change; it was not the first but was the biggest conference on climate change at the same (and remains one of the biggest of all time) (Khooshie and Panjabi 1993; Schreurs 2012). Many images that came out of the Rio Summit captured the world's attention, such as those of indigenous Amazonian people with feather headdresses demanding an end to deforestation. The Rio Summit attracted activists and NGOs, as well as heads of state. Most remarkably, the Rio Summit gave rise to no fewer than three UN environmental conventions, an unprecedented development in global environmental governance: the Convention on Biological Diversity (CBD), the UN Framework Convention on Climate Change (UNFCCC), and the UN Convention to Combat Desertification (UNCCD). This was a time when environmental crisis narratives were at their peak, and the UNFCCC was an explicit recognition that climate change is a dire issue that countries must act immediately to address on a global scale.

These three UN conventions continue to collate the best scientific data available and distribute reports that make it accessible. These conventions make progress by convening groups of scientists every few years at scientific meetings called SBSTA (Subsidiary Body for Scientific and Technological Advice); these are where the data from scientists working around the world are brought together, summarized and synthesized, and draft recommendations formulated for consideration by policy-makers (e.g., states or "parties" to the convention). Subsequently, at COP (Conference of the Parties) meetings, the science-policy interface reveals itself as scientists, NGOs, activists, and national delegations come together. Resolutions, articles, and amendments to the convention are negotiated through a grueling process of line-by-line word-smithing (bracketing), until language is agreed on for the final high-level segment in which senior government leaders converge (see, for example, Witter et al. 2015). One of the key jobs of the IPCC and these other conventions and intergovernmental bodies is to produce scenarios and provide recommendations about the various kinds of threats, risks, and vulnerabilities to different places (such as small island nations). There is a huge volume of data to boil down to tables and brief summaries for decision-makers who would not otherwise pay attention or know how to interpret it.

The Paris Agreement, which continued the work of its previous iteration known as the Kyoto Protocol, came out of the UNFCCC and is focused on securing global commitments to reducing greenhouse gas emissions. Almost every country in the world is a signatory to it, with the understanding that combating climate change must be a global effort, especially by high-emitting countries like the United States. In December 2015, President Barack Obama signed the agreement on behalf of the United States. In November 2020, President Donald Trump officially withdrew U.S. participation, fulfilling a campaign promise and working to further dismantle environmental regulations meant to safeguard the natural resources, including the air we all breathe, of the nation and the world. On his first day in office in January 2021, President Joe Biden fulfilled his own campaign promise to sign an executive order to reinstate it, which was made official on February 19, 2021.

An important part of climate change policy is recognition of the fact that the Global North is responsible for most of the climate change, while the effects are most directly felt in the Global South (Hurrell and Sengupta 2012). Brisman, South, and Walters (2018, 1) write: "The politics and conquests of the Global North have long necessitated the forced migration, colonization and ecological plunder of the Global South for imperial and capital expansionism," even using the term "climate apartheid" to call attention to unequal burdens of climate change on people of the Global South. These unequal burdens are predicated on legacies of colonialism, making it a contested and hugely consequential political debate. Vulnerability to the effects of climate change can be estimated, measured, indexed, and shown at different scales (Ford et al. 2018). One of the biggest conversations that has emerged in recent years is how the effects of climatic changes on rainfall, droughts, temperature extremes, and dangerous weather events produce even more vulnerability among already vulnerable populations, and how these changing conditions affect movement around the globe. Climate refugees are already migrating from Central America as a result of more frequent and intense hurricanes (in addition to violence and political instability).

As sea levels rise and extreme events devastate coastal areas with increasing frequency, the idea of managed retreat from places that are becoming uninhabitable links with concerns about environmental justice, both between countries and within countries in the Global North such as the United States. Climate change, like many other environmental, economic, and social challenges, has disproportional effects on communities of color in the United States, for example. The concern becomes which populations have assets that allow them to adapt to climate change, and which do not. We found that in the southeastern United States, in communities ranging from urban, suburban to rural, concern about climate change was often, but not always, roughly proportional to perceptions of personal vulnerability to effects of climate change (Himmelfarb et al. 2014).

PUBLIC UNDERSTANDINGS OF CLIMATE CHANGE

Current debates about climate have been studied by anthropologists and other social scientists who are uniquely positioned to provide nuanced understandings of how people who are not climate scientists conceptualize and accept—or not—the reality of climate change (Crate 2011; Crate and Nuttall 2009; Kempton et al. 1996). These scholars have examined the broader societal dynamics that influence local attitudes about climate change, such as social, political, ecological, economic, and even religious factors; understanding the interplay between these is a critical component of the social side of climate change research (Crate 2011; Roncoli et al. 2009). Vedwan and Rhoades (2001, 109) state that "understanding local perceptions of the relative amounts, direction, and impact of climate change are key to arriving at an understanding of patterns in human responses." To this we add that predicting and managing responses to climate change are dependent on understanding how perceptions are formed, articulated, and shared. To this end, the aim of this chapter is to provide context, as anthropologists, to what we heard at our field sites about climate change and its relevance to bioenergy. But first we need to provide a bit of background regarding major influences on the public's perception of climate change and its relation to bioenergy.

As we explored in previous chapters, there sometimes arise "moments of influence" (Witter et al. 2015) that lead to paradigm shifts in thinking. One such moment came in 1998 with the creation of the "hockey stick graph." Graphics are key to public understanding of the dangers posed by climate change, and while the data itself is apolitical, visual representations of this data are pulled into political debates on all sides. The "hockey stick" metaphor has had a lasting imprint on the minds of many regular people (i.e., not scientists) around the world. Wolfgang Behringer (2010, 3–4, referencing Mann, Bradley, and Hughes 1998) explains the origin of this metaphor:

> In the eye of the political hurricane were the originators of the "hockey stick theory": the climate researchers Michael Mann (Pennsylvania State University), Raymond S. Bradley (University of Massachusetts) and Malcom K. Hughes

(University of Arizona), who in 1998 presented a study on global warming in the past six hundred years. They claimed that on average the 1990s had been hotter than any other decade in the last six centuries, and that this global warming was "anthropogenous," that is, attributable to greenhouse gases produced by human activity. Their climate curve did not cause much surprise at first, as the longest stretch of the period in question had been marked by the global cooling of the Little Ice Age. But, shortly before the millennium celebrations, the climate researchers increased its scope by another four centuries. The climate curve of the last thousand years then acquired the form of a hockey stick: not much happened for nine hundred years, but then in the late twentieth century the temperature curve showed a rapid rise. The form of the hockey stick came to symbolize this way of seeing things.

This hockey stick metaphor has been called "the most controversial chart in science" (Mooney 2013) and has been the subject of intense criticism by climate change deniers, often led by industry leaders who deny their role in propelling it. Regardless, these efforts at denial have largely failed (though not necessarily on local levels in the rural South, as we shall soon see).

This concept has been recently repopularized by Michael Mann, a climatologist and geophysicist, and as noted, one of the originators of the metaphor, who rails against the ongoing assault on climate science (Mann 2012). He explains that there are multiple hockey sticks. These include changes over time in temperature, sea level rise, contraction of sea ice, and more. They all point to the same thing; the same general hockey stick-shaped curves. What, he asks—many of us ask—will it take to bend the curve back down? Mann (2012, 257) suggests:

Yes, the public discourse has been polluted now for decades by corporate-funded disinformation—not just with climate change, but with a host of health, environmental, and societal threats. Public perception is fickle. Things could change quickly with a concerted effort to improve the public's understanding of climate change risks and what's likely to be lost by not addressing them. Such an effort would of necessity require scientists to work closely with social scientists and with public policy and communication experts. It would require the financial support of foundations and private sector interests that genuinely care about Earth and its future. It would need to take full advantage of Internet-age organizing opportunities, using social media and online networking tools, to build a true grassroots movement that can go toe-to-toe with the massively-funded "Astroturf"[3] campaigns. And it would certainly need to work toward a dramatic improvement in the accuracy and objectivity of mainstream media coverage of the climate issue. Only an informed electorate can hold our policy makers accountable to represent our interests and values and insist on the development of a sensible climate strategy.

Mann, who was "cautiously optimistic" in 2012 that these changes might be brewing, also notes that as we grapple with understanding difficult and emotionally charged issues such as extinction and ecological collapse, it is easy to succumb to feelings of helplessness and hopelessness; his message is that it is (probably) not too late. In an interview in 2021 (Watts 2021), he said:

> I am optimistic about a favourable shift in the political wind. The youth climate movement has galvanised attention and re-centred the debate on intergenerational ethics. We are seeing a tipping point in public consciousness. That bodes well. There is still a viable way forward to avoid climate catastrophe.

In other words, we need to act, but we also need to not lose hope. He goes on to say that "Doom-mongering has overtaken denial as a threat and as a tactic . . . if people believe there is nothing you can do, they are led down a path of disengagement. They unwittingly do the bidding of fossil fuel interests by giving up."

CLIMATE CHANGE DENIAL

Narratives in opposition to climate change increased dramatically as the U.S. government was finalizing President Barack Obama's climate policy designed to limit and reduce greenhouse gas emissions from power plants through the Clean Power Plan, and government officials were also readying the country to join the Paris Accord at the United Nation's global climate summit later that year (Cama 2020). If ever there were two incommensurable perspectives, it would be between climate realists who accept the scientific consensus on climate change and climate change deniers who refuse to acknowledge the scientific consensus on climate change. We are currently living in an era where diverse and incommensurable narratives and imaginaries, some of which mischaracterize science or are inconsistent with the scientific consensus, are in wide circulation through highly polarized institutions and media.

Oreskes and Conway (2010), in their book *Merchants of Doubt*, review why the attack on science supporting climate change has largely succeeded in preventing substantive action on climate change at the highest political levels. It has done so by taking control of the narrative and taking hold in the public arena, popular media, and even some researchers within the scientific community. The authors document how climate change denial can be seen to be rooted in lessons learned from Big Tobacco, who established a dominant narrative that the science on the links between smoking and cancer was unsettled; that cancer is a mysterious disease with many causes; that the industry provides rural communities with a number of high-paying jobs; and that restricting smoking, or even discouraging it, constitutes an assault on American freedom. These are seen as strategic maneuvers designed to produce skepticism and doubt, and climate change deniers have pulled liberally from these scripts (Shepherd 2011).

Climate change denial promotes narratives based on incomplete understandings of, or outright disdain for, science. Climate change deniers misinterpret the ways that climate scientists talk about uncertainty, and they misinterpret the range of possible scenarios as a lack of confidence in the models themselves. They do not acknowledge that scientists are aware of naturally occurring fluctuations and are able to incorporate these into their models, separating them from anthropogenic causes for specific changes in temperature or atmospheric levels of carbon or methane. They ignore scientific understanding that effects of climate change are uneven, and that different areas will experience different effects, or perhaps no effects. They do not account for the ways that some populations are more vulnerable to climate change, or that climate change is expected to exacerbate their vulnerability. They do not see connections between changes in climate and human migration patterns, intensified storms, or new disease outbreaks. But they provide a simple and powerful message that casts doubt on the existence and causes of climate change, a message that has been widely circulated in the media and public discourse.

Public opinion surveys over several decades find that while public awareness and knowledge of climate change have increased, more Americans believe that climate change is occurring than believe that it is caused by human activities or have a high level of concern about it (Schwom et al. 2015). Notably, for the past two decades, the United States has also had a not-insignificant segment of the population holding climate change denial beliefs (Schwom et al. 2015). As noted above, there has been an organized climate change denial movement driven by powerful interests that has perpetuated a common understanding of uncertainty about the science (not a scientific understanding of uncertainty), while aligning with societal pro-economic growth and anti-environmental messages (Dunlap and McCright 2015). In other words, climate change denial has muddied the waters, cast doubt on science, and actively made it a political rather than a scientific issue. Discourses rejecting climate change invoke various framings and linguistic elements, related to the goals of originators, intentionally driving disagreement about climate change (Hulme 2009).

CLIMATE CHANGE IMAGINARIES

Social processes are involved in the interpretation of science at every step, and climate change is no exception (Malone 2009). As noted previously, language is an important component of imaginaries and discourses, and this is particularly true in the climate change science and policy conversation. Several key terms, which we reference in our own research, are cited time and again by climate researchers and policy-makers. First, the term "adaptation" is itself an admission that climate change is happening, and that we must develop mechanisms by which to adapt to it. These adaptations, which can occur simultaneously on multiple levels, can entail interventions in architecture, manufacturing, farming, urban planning, legal strategies, and many other realms of daily life. Second, the term "mitigation" addresses the question of what can be done to stop or slow climate change and its deleterious effects. It addresses the question of how we can bend the curve back to a manageable level, and what we can do to lessen the human impact on climate. Third, "vulnerability" highlights the fact that climate change affects human lives differentially. Researchers have proposed a number of different indicators of vulnerability, while policy-makers must decide on specific measures to implement in order to reduce known and potential threats and risks to particularly vulnerable populations.

Malone (2009) lists eleven different "argument families" that are used when talking about climate change. Adapting these to the bioenergy context, these include some of the basic climate change denial arguments—such as "scientists are incorrect and climate isn't changing," or "climate is changing but not because of human activity"—that take climate out of the argument altogether. They also include a view of climate change as a tool of the rich and powerful to take advantage of ordinary people. On the environmentalist side, another one sees climate change as reflecting humanity's broken relationship with nature; utilizing forests to address it only magnifies the problem. However, in the middle are a series of arguments about the need to use the tools of modernity, such as technology, policies, and markets (Malone 2009); these correspond quite closely to sociotechnical imaginaries. These arguments are often linked together in discourses and imaginaries that inform the way people talk about climate change and bioenergy, as we shall see in the excerpts and discussion of our interviews.

WOOD-BASED BIOENERGY AND CLIMATE CHANGE

The impact of wood-based bioenergy on climate change is complicated, and evaluating it often depends on complex modeling. There have been numerous analyses with different conditions and assumptions that have produced different results. An influential early analysis focusing on Massachusetts noted that the greenhouse gas implications of burning woody biomass depend on many factors, including how the bioenergy is used, what fossil fuel it replaces, the nature of the forest management system under which it is produced, the rate of regrowth of forests, and the rate of natural disturbances (Walker et al. 2010). By analyzing the immediate carbon debt, greenhouse gas emissions from burning biomass, and the gradually accruing carbon dividends, they find a lag time for carbon debts to be repaid range from five to ninety years. This finding suggests significant lag times for climate change benefits to accrue from burning woody biomass, while at the same time noting the need for actions to ensure sustainable forest management (Walker et al. 2010). A different study in the U.S. South found positive climate benefits for woody biomass use from reforested pine lands, while also emphasizing the need to test additional forest management practices and for appropriate policy mechanisms and sustainable forestry (Dwivedi et al. 2016). Other studies argue that policies and extension efforts can ensure climate benefits from bioenergy while also improving local livelihoods (Mendes Souza et al. 2017), and that bioenergy with carbon capture and storage (BECCS) can have considerable climate benefits (Creutzig et al. 2015). While there are certainly many additional studies, these serve as illustrative examples of the complexity of the relationships between analyses of bioenergy and climate change, as well as to highlight the importance of local forestry practices, policies, and technologies.

These studies provide important context for the opinions that we heard in our field sites about climate change and its relation to bioenergy. As discussed in the previous chapter, many people in the U.S. South have deep attachments to trees and forests, and many people in rural areas have both sentimental and pragmatic approaches to valuing and managing forests. For many landowners and community members in rural areas, especially in what are—or were— forest-dependent communities, the production of woody biomass for bioenergy provides another market for wood products. The development of this market has the potential to revitalize the forest-based economy and encourage the planting of more trees instead of large-scale conversion to development or agriculture. As we will show, even rural landowners and owners of wood products industries who are not committed to the idea of anthropogenic climate change understand that climate change is a major driver of the development of new wood-based bioenergy facilities and technologies. They can see that the production of pellets or liquid fuels from wood "differs from preexisting forest product industries in that its legitimacy derives principally from its contribution to climate change mitigation" (Palmer 2021, 141). Belief in that legitimacy is not universally shared; as we will show, both climate change deniers and environmental advocates who claim that wood-based bioenergy will exacerbate rather than mitigate climate change question this assumption. However, the ways that different actors value—or don't value—the science of climate change have direct effects on their opinions about bioenergy.

PERCEPTIONS OF CLIMATE CHANGE
AND BIOENERGY IN THE SOUTH

Mitigation of climate change is fundamental to the national bioenergy imaginary in that it provides an overarching public goal in support of bioenergy development. It coexists with rural development and energy independence, but climate change and renewable energy provide the environmental justification. Yet climate change, especially when it places the blame on human activity, does not provide a major justification in the South—even when people note the changes they have experienced firsthand. As one extension forester said, "Climate change, you have a hard time in Mississippi saying it's manmade. It's a hard sell for people to understand it. But they've got to see the anomalies in the weather." Another forester discussed the declining populations of rattlesnakes in southern Georgia, who have a symbiotic relationship with endangered gopher tortoises whose native habitat is in established longleaf-dominated pine forests:

> The tortoise . . . there are some around here interested in saving the gopher tortoise. Those gophers are getting gone. But people don't necessarily kill them now like they did 20 years ago. It's more contentious now. Maybe there's just less rednecks [laughs], that go out and spotlight deer . . . My theory is, to protect the gopher tortoise and the rattlesnake . . . it's a natural cycle. But extinction blows our minds . . . This area is still rural. I can't say development has been the ruin of the tortoise and the rattlesnake. It's not logging. You may run over a few here and there, but in the big scheme of things . . . that's not making them extinct. It could be a tweak in the climate . . . Snakes are much rarer than they used to be.

Note that he is not totally committed to causation here, only correlation; however, his mind is open to the possibility that the ecological changes he sees are part of a larger pattern of climate changes.

Some of these personal observations of climate change are accompanied by alternative explanations that are not supported by science. For example, one Black community leader discussed a case in which another community member suffered a snakebite at a time when snakes should still be hibernating. In noting changes in snake behavior, which he attributes to climate change, he then suggested that climate change is caused by satellites:

> [*Name*] got bitten by a rattlesnake in January—there's been changes, not like when you grow up . . . Global warming . . . all those satellites, they're changing Mother Nature's thinking. Technology just goes advancing itself. I appreciate technology, but it goes a bit much.

We should add here that in previous research that we have done in urban, suburban, and rural Georgia focused on perceptions of climate change and the relationship between social vulnerability and climate change (Johnson Gaither et al. 2016; Himmelfarb et al. 2014; Schelhas, Hitchner, and Johnson 2012), we heard several people (of different races) mention the purported links between climate change and satellites as being the culprit behind changes in the Earth's atmosphere.

A number of people expressed the view that climate change is not a daily concern, and much of the indifference seems to be rooted in what they see as uncertainty among climate scientists (which again is not an accurate reflection of the scientific definition of uncertainty). One forest landowner laughed while telling us, "You put global warming to a vote, and I'm all for it. Some people think they know the answers—maybe they do and maybe they don't, about what causes changes." Other people, such as bioenergy industry representatives and natural resources professionals, recognized the lack of understanding about the science of climate change and said that they tried to educate people in certain situations. For example, one forestry professional in Mississippi is working with clients to help them prepare their forests for coming changes in climate, saying:

> Climate models show that the southeast will be cooler and wetter. The rest of the area will be hotter and drier. Droughts and floods will be the new normal . . . There will be changes. The hurricane predictions this year didn't pan out at all. No . . . they were right, they just didn't make landfall. There are huge economic impacts when it hits land.

He understands the uncertainty inherent within climate science, and the parallel with forecasted hurricane trajectories, and felt that it was important for forest owners to understand how climate change might affect their forests over time. However, many people misunderstand the scientific definition of uncertainty and think that scientists just throw out wild predictions that sometimes come true. Other people indicated belief in climate change but could not articulate how it works. One farmer in Georgia said, "You gotta have trees to clean your carbons and stuff like that."

While in many interviews, climate change was not mentioned at all, we found climate change was discussed more frequently by speakers at bioenergy conferences, where it was invoked in several different ways. One speaker with a more environmentalist orientation said at a conference, "Climate change is upon us. We'll continue to see support for bioenergy as a tool for mitigation of climate change . . . We need negative emissions, not just [carbon neutral emissions]." This speaker accepts climate change, supports bioenergy, and emphasizes a focus on renewability. The same speaker, at another conference, emphasized his support for forestry as a way of providing returns from working forests and thereby reducing the incentive to clear forests for development, fitting wood-based bioenergy into a sustainable forestry frame. The contrast in his two sets of remarks also demonstrates that at one conference, he felt compelled to include climate change in the conversation, while in another, he left it out to focus on forest sustainability. These different framings also highlight the difference in perspective on climate change between bioenergy producers in the United States such as the Georgia Biomass pellet mill in Waycross, and bioenergy users in Europe. In other words, if the Europeans believe in climate change, then bioenergy producers in the United States should promote that aspect, even if climate change is not as important to the producers; they can instead take pride in their work of improving trees and increasing forest cover. If you believe in both, as this speaker does, then wood-based bioenergy produced in an ecologically sustainable way is a productive path forward for all sides. A researcher in Louisiana, noting how climate was not a major component of the discourse about bioenergy in the South, said, "Climate is not part of the discussion. No one's discussing the whole cradle to grave." In other words, he said, some actors pushing the pro-bioenergy imaginary only calculate part of the equation; the rest of the math is just not there. We find this analogous to cutting off the hockey stick at the end of the shaft before the curve of the blade begins. As noted, even non-scientists are not always fooled by this sleight of hand. We will come back to the skepticism, from a diversity of speakers, that arises from presenting wood-based bioenergy as a potential solution to climate change.

Markers of Sustainability

Sustainability and renewability discussions lead directly into the use of forest certification programs designed to ensure that these are a part of woody biomass production, such as the Sustainable Forestry Initiative (SFI) and the Forest Stewardship Council (FSC). Certification may be required to sell forest products in certain markets, generally those that are highly visible to a public with environmental concerns. A timber company official noted that the European pellet market requires certification: "All of our certification can be traced back to Europe. We have SFI fiber sourcing. Just because of pellets; we don't need it for lumber."

However, discussions about certification often devalue issues of renewability and sustainability, when they meet resistance from forest owners and managers who do not welcome the intrusion into their affairs and who question the rationale for sustainability in the first place. A natural resource professional said, "There's a pellet mill there. It's had a huge economic impact on the area. I don't know how they claim to be renewable, shipping pellets to Europe. With that government, they're forced to do it." A forest landowner suggested that even when certification is required, it was implemented by forest product companies in a haphazard fashion and only done to, as mentioned earlier, "to cover their ass." These quotes raise questions about the certification process, and carbon neutrality in particular. Though certification is required by European markets, forest management professionals in the United States express doubt about the legitimacy of the process, in part because renewability seems unlikely to them when clear figures have not been shared. They feel they are not given enough information by buyer companies or by governments to determine the overall sustainability of the bioenergy system. They also suggest that companies only obtain certification in order to meet the market requirement; it is, in essence, meaningless in an ecological sense. One forester called certification "hogwash." But then he went on to say that it may be needed anyway, acknowledging that most members of the public perceive that forests with certification are being managed more sustainably.

Ultimately, those involved in the forest products industry tended to see certification as inevitable. One person told us that it was imperative that landowners be able to demonstrate sustainability, however much they feel that certification is unnecessary and onerous. If they refuse to participate in certification processes, "the industry will pass them by." This is a pragmatic, free market-based perspective stating the need for certification. But these statements from our research reveal the different perspectives on certification and its legitimacy across scales, something that ultimately will have to be addressed in larger efforts to use bioenergy to address climate change.

Interestingly, southern foresters and landowners generally view forest management as inherently sustainable and trees as a renewable resource, and they tend to resist market-based approaches like forest certification for sustainability. The tendency has been for people to see it as outside interference in their forest management, which they believe was already sustainable. They feel they already do everything required of them with certification, yet they also have to pay a great deal of money and submit to external audits. They feel they are being forced to adhere to regulations that do nothing to improve forest health, only appease those with power and line their pockets. As noted, they are not fond of having "new regulations shoved down their throat." Here, even a market-based approach is equated with government regulation. As such, both climate change and certification were often seen as a political and not a scientific issue.

Given that certification of forests for bioenergy is seen as an international requirement, landowners compare the U.S. South to other countries, arguing that forests are increasing in the southern United States when they are being cleared in other countries. One landowner mentioned rainforest destruction and beef cattle in the Amazon, a topic that has received considerable public attention in the United States, saying, "They don't replant there. In Georgia, there's a 31- or 32-year cycle, and then it starts over. There are more trees now than there used to be." The managed forest sector in the U.S. South does generally involve replanting after harvest, but reducing the issue of forest health to a measure of tree cover misses many aspects of sustainability and renewability.

In some instances, people talked about climate change almost in passing, as an afterthought. For example, one forester said that cutting small-diameter trees for bioenergy could be a way of reducing fire risk, which had increased with drought that "some would say [was due to] climate change." He expressed his doubts, but acknowledged the possibility of climate change, and the potential role that bioenergy development could play in reducing its impacts in the future. In any case, renewability and sustainability have long been claimed by forestry and forest owners, although their interpretation of these terms tends to have a narrow focus. These terms may refer only to the continued production of wood products, may be expanded to include wildlife (both game and non-game species), or may more broadly approach national definitions and certification programs.

We found that people in the southern United States generally assume that forestry is a sustainable way to grow a renewable resource. Landowners and community members, as well as those in the forest products business, noted that trees are renewable and that the industry has been sustainable for a long time. Another landowner said, "I'm for the biomass. It's a natural resource. It grows again, not like coal, gold." Similarly, a forest owner noted that trees are a renewable resource, and that he would like to see a bioenergy market develop so that more of each tree could be utilized; he said, "We hope that we can use the whole tree. There is so much waste when all we can do is burn it or let it rot." From his perspective, the focus of forestry is on forest products alone, and anything left over after thinning or harvest is wasted. However, one forester emphasized his attention to the larger forest, viewing it more as a functioning ecosystem, rather than just a source of merchantable products: "I'm a forest manager, not a tree manager . . . I like healthy forests . . . In a forest, it all has its place." But whether trees are seen as a commodity or as vital parts of a woodland ecosystem, sustainability can be invoked with or without mentioning, much less prioritizing, climate change.

Regardless of their feelings about certification requirements, the long-term sustainability of the forest industry was the main concern to foresters, landowners, natural resource professionals, and bioenergy industries and advocates. What we heard from a number of forest landowners, foresters, and forest managers was that while there is currently enough wood to supply bioenergy markets as they exist now, expansion of these markets could eventually strain the forest resources of the southeast. One forester in Waycross explained that at the time, the Georgia Biomass facility was taking about 250 loads of wood a day, a similar amount to a medium-sized pulp mill. Georgia Biomass, he explained is, "a player, but not the big man on campus. Pulp mills are still the biggest. They take 500 loads a day . . . they haul from here to Brunswick." However, he said that if the pellet mill required the same amount, it could present a problem:

> All the pellets are exported. That goes through my mind, not that it's going to Germany, but away from here. The question I hear a lot, and I don't believe it, but if they're taking 250 loads a day, where's it coming from? All the wood will be gone, they say. But you can see it everywhere. It's that rape and pillage mentality, but you're just not seeing it. If they take 500 [loads a day], that may be too much.

This measure of, and concern for, availability of biomass shows that the forest industry and environmentalists both realize that trees are a finite resource; however, when it comes to harvesting them, they often disagree on how much is too much.

Hostility Toward Climate Science

While many people with whom we spoke were on the fence about climate change, some were openly hostile to climate change, invoking anti-environmental and pro-fossil fuel narratives. Several people took the opportunity to declare their support for the Keystone Pipeline and fracking in gas fields, or coal, suggesting that we just need to capture emissions (thus invoking the "technological fix" narrative). This was mostly an anti-environmentalist argument, as one person claimed these initiatives were being obstructed by "the Department of Environmentalists." One community member expressed belief that climate change is totally fabricated, citing the "ignorance" of people who have been influenced by climate scientists (and the famous "hockey stick graph"):

> Global warming has been found to be fraudulent, based on fraudulent information. How they've used these fake numbers and that hockey stick graph. They figured, "No one will catch it, we'll just put it in there." That just shows the ignorance of 90% of the people. They'll go home, and the only news they'll hear is the 6:00 news, which is biased towards things happening in this county.

They won't discuss the truth. A whole lot of people have had it up to their ears. This next election is the most important one we've ever had. A lot of people don't vote or don't become educated about what's really happening, changes in government, rampant spending . . . Everything is connected.

This speaker clearly states that climate change is a political, rather than scientific issue, blaming President Barack Obama for perpetuating the myth of climate change.

Blaming the Environmentalists

When climate change is viewed as an environmental issue, environmentalists are dismissed along with it. They are derided for interfering in other people's business, not doing their part, and using their elite status to promote NIMBYism. This point hearkens back to the farmer we quoted in the last chapter who was furious that environmentalists criticized him for harvesting his trees when "they've never planted a tree in their lives." Environmentalists are often also framed as talking in abstractions that have no basis in lived experience or in observable reality. To this point, a farmer in Mississippi said:

Environmentalists talk in generalities. You can't put generalities in your truck. Those environmentalists—they say they want it, but you can't put solar, wind, nuclear, corn dust in their backyard. They're all for it, but not if they have to deal with the consequences.

In other words, people who promote environmentalism should be willing to shoulder some of the associated burdens of it, while those live close to the land, including farmers and forest landowners, need concrete solutions to observable problems. An extension forestry professor said, "Landowners can't go along with 'the sky is falling.' That won't work." Rather, they need to develop specific actions that make sense for them, putting the focus on their own economic activities and landholdings rather than broader national and global concerns. If planting and tending trees and crops that help them earn a living have the side effect of mitigating climate change, so much the better; they would be doing it anyway without input from environmentalists.

Another community member who was particularly outraged about the claim that global warming was an important environmental issue dipped into the tax issue and said:

I almost get sick when I hear the word "green." The "greens" generate a lot of self-serving information such as global warming. The only reason global warming has been such a hot topic is because Al Gore has been making millions of

dollars out of the word "green."[4] And he's one of the biggest abusers of energy
. . . Carbon emissions . . . that's a big farce. . . Trees use carbon dioxide . . . .
Planting trees is one of the best ways . . . But carbon emissions are not critical,
not real. . . . They want to charge the airlines a carbon usage tax. Who's going
to tax China and India?

What is notable here is the mixing of multiple climate change denial narratives, including strong anti-environmentalism, anti-globalism, anti-science,
and anti-tax elements. The speaker *almost* admitted that trees store carbon,
but quickly declared that unimportant and reverted to the series of statements
drawn from climate change denial narratives. That this comes up in the
context of talking about bioenergy reveals the prevalence of climate change
denial and the way that climate change is deemed unimportant to bioenergy
development at the local level in the communities where we worked.

Environmentalist Perspectives

Though often discounted, we found that environmental advocates enter the
bioenergy conversation in multiple ways. As discussed in the book's introduction, some environmental groups active in the southeastern United States,
such as the Dogwood Alliance, are vehemently opposed to large-scale wood-
based bioenergy production, particularly products like pellets that are shipped
overseas. Others are, again, seeking that balance between the potential for
wood-based bioenergy to help mitigate climate change and its potential to
exacerbate it. But despite differences such as these, environmental organizations are not shy about talking about the importance of climate change,
regardless of audience. However, their messaging, at least at the regional
bioenergy conferences we attended, was intended to find productive ways
forward instead of antagonizing. For example, a representative of the environmental organization SACE (Southern Alliance for Clean Energy) stated at
one conference in Georgia:

> Climate change is upon us. We'll continue to see support for bioenergy as a tool
> for mitigation of climate change . . . Sustainable bioenergy is indispensable for
> climate mitigation—we need negative net emissions, going beyond neutral. Not
> all bioenergy is climate friendly. Wood pellets are CO_2 intensive, when trans-
> port across the ocean is included.

He then presented a chart detailing the climate benefit of bioenergy, which
put large-scale pellet production in the transitional area between carbon
neutral and carbon positive (providing a range of scenarios dependent on
different variables, as scientists do). Advocates of wood-based bioenergy,

including a number of foresters and forest products industries, recognize and appreciate the differences between some of the environmental NGOs. One state forester in Georgia told us:

> Dogwood Alliance, Sierra Club, NRDC [Natural Resources Defense Council], none of them want to do anything. We work with conservation groups. Some are proponents of biomass. The biggest is the Southern Alliance for Clean Energy (SACE). Some environmental groups would support it, but they're very skeptical, very afraid to do something that's not right. So there's an educational hurdle. They have no background in farming, conservation.

The view that environmentalists generally don't know what they're talking about was quite prevalent among many forest professionals, forest landowners, farmers, and other rural community members. To the extent that they felt like their conversations could be productive, they felt a duty to give environmentalists a dose of reality and inform them about life on the ground. Environmentalists, of course, expressed equal certainty in the rightness of their perspective.

Making Space for Climate Change Denial

Regardless of these debates and the existence of denial, climate change remains an important justification for public support for biofuels in the form of policies and incentives. We noted that speakers at local workshops often hedged their acknowledgement of climate change to account for the prevalence of climate change skepticism. Such adjustments of discourse were likely rooted in a speaker's knowledge that climate change skepticism is common in the rural South and the forestry profession. They reflect an attempt to connect with this audience as bioenergy is promoted at the local level and to prevent it from being too entangled with climate change. This type of equivocation, which essentially renders climate change irrelevant to the need to pursue bioenergy development, was evident in the presentations of some speakers at workshops and regional conferences we attended. One bioenergy company representative said in his talk:

> [*Our company*] is really special in that it leads to an 80% reduction in greenhouse gas emissions. Now, global warming . . . they say it's about carbon that was stored a long time ago, burned, and put back in the atmosphere. You ask 10 scientists about it, and you get 11 different opinions. But the fact is that trees grow—they store carbon and release oxygen. [*Company name*] takes that carbon, makes fuel, emits CO_2, which trees absorb. It sustains the life cycle of carbon.

This speaker did not blatantly attempt to sidestep the issue of climate change; he claimed that emitting CO_2 was a good thing because it helps trees and "sustains the life cycle of carbon."

There are other people who operate in two worlds with radically different—even incommensurable—ways of talking about climate change; they find ways to talk about CO_2 and carbon while also signaling uncertainty about climate change. Even if they admit to climate change, they say it is not enough to get people's attention. One workshop speaker stated:

> There is a need for actions; combating climate change, jobs, affordability of energy and living, security of energy supply, defend the long-term competitiveness of businesses, help the agriculture and forestry sector. This brings power to this whole debate. Just climate change doesn't make people do something.

He thus put climate change front and center, but also recognized that it was not always the primary driver of action here in the United States. This speaker also tailored his message to the audience, giving them advice on how to tailor their messages to those who deny or are unconcerned with climate change. He takes climate change as a fact, but acknowledges that not everyone does.

Questioning the Role of Bioenergy in Climate Change Mitigation

As mentioned earlier, other people, even while skeptical of climate change in general, found it difficult to believe that bioenergy was really a legitimate way to address it, given environmental costs associated with transporting bioenergy products. This was especially evident in discussions about shipping wood pellets from the United States to Europe, as noted by diverse stakeholders. One community leader, who clearly does not take climate change as a given, also questions the coal-based energy that is used to produce pellets and the emissions from shipping them across the Atlantic Ocean:

> Let me ask you a question. So they bring the trees in on a logging truck, they make these pellets here and they get their power from Georgia Power, which is run off coal, then they run it on a train and then ship it across the ocean, and it's still coming out carbon negative? I mean, *if* you believe in global warming . . . Until it's as cost-effective as coal, [wood-based bioenergy] can't compete unless it's driven by politics. The market is political.

A community member also questioned the renewability of bioenergy products shipped overseas, laughing when he said, "I think about the sustainability of it—is it still renewable after all the transport? Maybe we'd be better off digging coal in Russia." In contesting bioenergy development as a means to address climate change, these quotes reflect the underlying local notion that bioenergy and environmental regulations requiring it are part of a government overreach or a scam for people to enrich themselves with little public benefit. They also often reflect the continued support for fossil fuels, and perhaps a lot of resistance to change—particularly change driven by environmentalists and the government.

Distrust of Carbon Markets

Carbon markets were widely panned; most people with whom we spoke doubted their efficacy at best and were vehemently opposed to them at worst. People often described climate change as an issue fabricated to advance other agendas or for financial gain; carbon markets, from their perspective, were proof of this. One community member stated, "There is no global warming . . . Those people want that carbon tax, so they say they believe in global warming." A landowner said, "I don't buy into carbon sequestration [carbon markets]. If you just give [something] to me, it scares me." Another discussed what he had personally witnessed with carbon credits:

> There was a meeting at the Citizens Bank to discuss carbon credits about 2004. It was full of landowners. They hired locals to sell the idea. A lot of people signed up as mid-brokers, mostly consulting foresters . . . Carbon credits have died away. That was a dog and pony show. I'd hate to know how much the government spent on that. A lot of people jumped on that. You had to leave your stand for a certain period of time and restrict harvests, and your management had to be okay . . . Some got money at first, then the brokers came in to trade and sell credits. I don't believe carbon credits lasted two years.

Comments like these touch on anti-government and anti-environmentalist sentiments, while at the same time expressing distrust in what is seen as "artificially created markets" that result from international agreements and from U.S. policies that could easily be abandoned or changed. A community member in Mississippi looked for a technological fix instead, saying, "We've got plenty of coal. We need to learn to capture those [emissions], need to learn to use that. I'm against carbon tax credits. That's a no-no. That will destroy our country."

Here we see another framing of bioenergy, and the environmental regulations it entails, as a solution to a problem (that's likely not even really a problem) more easily fixable by improvements to known technologies and products. After all, why bother fixing what's not really broken in the first place? Those that want to, it seems, are in it for some other reason. This skepticism about government officials, environmentalists, climate scientists, and others pervades the thinking of many rural Southerners. As we have noted, these opinions are shaped by myriad factors and circulated among people in various, overlapping social circles.

THE INTERSECTIONS OF CLIMATE CHANGE
WITH THE BIOENERGY IMAGINARY

We opened this chapter with a quote claiming that bioenergy is a good thing whether you believe in climate change or not. From the rural population's perspective, where climate change skepticism is widespread, the association of bioenergy with climate change often is explicitly severed; bioenergy represents just another wood market for forestry as usual. Foresters and people in the wood products industry see that bioenergy development is a response to climate change, and they recognize that wood markets need to demonstrate sustainability even though they think this is unnecessary. Environmental activists, when not outright opposed to using forests as feedstocks for energy, only support bioenergy when it is carbon neutral and sustains forests.

Sociotechnical imaginaries are, as Jasanoff and Kim (2015) suggest, "dreamscapes of modernity." They are put forth by powerful actors in an effort to change the future, both technical and social, and change is brought about with modern tools like technology, policies, and markets. Climate change is included in the bioenergy imaginary as one of the problems to be solved through bioenergy, which then justifies policies and incentives to promote bioenergy development. But climate change discourse is bifurcated; there are narratives that promote the urgency of climate change and counternarratives denying climate change. The tools of modernity are the bread and butter of government, large industry, and finance. National and global institutions work together in ways that address public goods and shape markets. But there are also other, widely circulating discourses that oppose the tools of modernity in the United States, which are often characterized by opposition to taxes, government, and science while in favor of an almost mythical free market state. While imaginaries that originate from powerful actors can be remarkably coherent and consistent, when they circulate among the general public, they encounter both sides of climate change and modernity discourses. These powerful imaginaries often originate at the national scale, but when they encounter the local scale where climate change denial is widespread and the tools of modernity are spurned—at least in oppositional discourses—the imaginary is altered or its impact reduced. This process is nowhere more evident than in the ways that mediating actors blend and hedge both sides of these discourses into the bioenergy sociotechnical imaginary as they bring it to the local level of implementation. Understanding the process by which this happens, including the ways it gets translated through discourses and across scales, is important for any effort to promote any sort of energy transition.

What does become apparent when looking at a range of studies is that bioenergy as a climate change mitigation strategy often boils down to an issue of scale. For example, removing large-scale transportation, the climate equation often shifts, a fact pointed out by a number of people with whom we spoke. So, some have proposed, having more smaller-scale bioenergy facilities in heavily forested landscapes, and distributing those energy products on a regional scale, for instance, instead of shipping them to Europe, would do more to mitigate climate change (as well as promote rural development, strengthen local wood markets, and contribute to national independence).

We have found that in both professional and everyday contexts, the complexities of science are often boiled down into more distinct messages and structured by discourses and imaginaries that shape public opinion and, often, management and policy. As we saw in the opening of this chapter, these messages are tailored to specific audiences in order to persuade; this process of strategic translation is perhaps more evident in discussions of climate change than any other issue we encountered. Again, we see how these issues are not discussed in isolation but rather as embedded in a matrix of other political, economic, and cultural values.

As we discuss further in the next chapter, some communities (and subpopulations within communities) are more vulnerable than others to a number of externalities of energy consumption and production. Environmental justice concerns in the siting of wood-based bioenergy facilities (and other producers of renewable energy) can be in direct conflict with energy justice concerns that aim to ensure equitable access to energy. As noted, communities of color often bear the disproportionate double burden of negative consequences of renewable energy development and climate change. Solutions that simultaneously address both, however, are not clear and simple. Bailey and Darkal (2018, 337) note that

> climate, energy and environmental justice share similar philosophies about fairness in the allocation of rights and responsibilities, but their different foci make it problematic to assume a neat alignment of goals. Our main interest here is intersections between energy justice and local environmental-social justice, because although climate justice forms a general backcloth to energy policy, energy and environmental-social justice often compete more directly with each other in renewable energy siting disputes.

What this often translates to on the ground is a series of messy conversations about which group's overarching concerns are prioritized at the policy level, as well as in development decisions in local communities. Efforts to accomplish all the goals at once, as the main bioenergy imaginary dictates, have so far met limited success. Still, the fact that climate science has proven that the global ecological system is at a "critical juncture point," the need to decarbonize the global economy remains paramount. Van de Graaf and Sovacool (2020, 229) note that "by far the most unifying force in global energy systems today is the challenge of climate change," leading to inter-dependencies between nations and a deeper understanding of the interconnectedness between energy and other issues. We aim here to demonstrate the interconnectedness between the issues of energy, landscape, climate, and race in our field sites in the U.S. South; however, we also call attention to the gaps which occur when imaginaries collide, scales don't match up, and values are ultimately incommensurable.

NOTES

1. https://www.ipcc.ch/about/structure/
2. https://www.ipcc.ch/about/preparingreports/
3. The metaphor of "astroturf" as used here refers to campaigns or organizations that pose as "grassroots" but are actually funded by corporations for the purposes of promoting an industry's financial interests.
4. As a popular and powerful figure, Al Gore draws a lot of ire. A natural resource professional told us, "people think Al Gore invented the internet. And global warming too."

Chapter 5

"The South Be the South"

How Bioenergy Development Illustrates and Affects Racial Dynamics in the U.S. South

Racial dynamics are deeply embedded in the landscape of the southeastern United States, and they directly affect how bioenergy imaginaries play out on the ground in rural southern towns. The cultural, social, and economic effects of the removal (and in many cases erasure) of Native American groups and the enslavement of Africans and persistent inequalities that their descendants (and other ethnic minorities) continue to face reverberate throughout the region. Like all habitable ecoregions, the southeastern United States is a palimpsest of many layers of human history, consisting of numerous and overlapping cultural influences, historical events and movements, and economic systems whose influence reaches far beyond its geophysical boundaries. The landscape has undergone numerous changes throughout human history, from land and resource use by pre-Columbian peoples to complex land management structures by various Native American tribes, to the arrival of European settlers whose subsequent conquest of the New World led to the "plantation kingdom" based on large-scale monoculture of such crops as cotton, rice, sugar cane, and tobacco (Follett et al. 2016). People taken from Africa and enslaved on these plantations were intricately tied to the land through both forced agricultural labor and subsistence gardening to supplement meager rations, and these ancestral experiences continue to influence the values, goals, and identities of Black landowners today. The dynamics of slavery are still written into the physical and social landscape of the South, and though the practice of slavery has been abolished, racial discrimination and disparities affect the everyday lives of minorities. The southern United States has a long, complicated, painful history with the issue of race, and while many people are willing to speak openly about it, it is often what is left unstated that tells us the most.

Race permeates discussions about bioenergy development in both explicit and implicit ways. The history of racial violence, dispossession, and disparities in the region directly affects differential valuations of forests, development policies, and climate change, as well as differential framings of social and economic issues and visions of a sustainable and just future. Racial concerns are an important and often overlooked aspect of wood-based bioenergy development specifically and forest-based rural development in general. The ethnographic analysis we present here reveals real and perceived impacts of bioenergy development in rural and urban communities, forcing the question of who bears the costs and who receives the benefits of bioenergy industry development. We encountered racial concerns regarding unequal hiring of minorities in local communities with bioenergy plants; siting of bioenergy facilities near minority communities causing concern over air quality and potential safety threats due to explosions or dangerous traffic situations; and a disproportionate level of insecure land tenure among Black landowners, restricting access to governmental assistance programs, forestry services, and opportunities to benefit from strengthened wood markets. Most importantly, we witnessed a relative lack of racial diversity among people with the power to make decisions about where and how bioenergy development is implemented in rural communities.

Though the region is a cultural mosaic representing a number of nationalities and ethnic backgrounds, in this chapter we mainly focus on the racial dynamics between White and Black community members and on the perspectives of Black community members with whom we have interacted. We are drawing from around 200 interviews, in addition to less formal interactions, with Black Americans in the southeastern United States that we have conducted as part of several different research projects, including ones focused on bioenergy, minority forest management and land title issues, and social vulnerability and climate change. We found that, not surprisingly, people vary widely in their abilities to see—and to articulate—the connections between historical racism, ongoing racism, and access to natural resources, financial capital, and new economic opportunities. While it would take many books to deeply examine the nuances of race in the southeastern United States, we are limiting our exploration of this topic in this chapter to specific subcategories of this topic that emerged from our research.

We begin with a general overview of the ways that people (of all races) talked about the history of racial relations in their communities, specifically as related to the history of forestry and forest products. While this varies by location (e.g., southern Georgia has a particularly long history with Black labor in the turpentine and naval stores industries), Black Americans have historically been involved with the most dangerous and labor-intensive aspects of logging throughout the South. This accumulated knowledge about

forests and forest products among some Black families is often overlooked
by mainstream foresters, as the forest management techniques that have historically worked best for Black landowners facing racism and lack of capital
(such as natural regeneration with low upfront economic inputs) do not fit the
economic models premised on intensive monocultural forest management.
Our research shows (and as we will see, many people told us this directly)
that the overt racism inherent in times of slavery and the Jim Crow era bleeds
into the present, where in many small, rural communities in the South, there
is little racial and ethnic diversity in the makeup of local decision-makers
(county commissioners, city councils, development authority boards, city
planners, etc.). This was not universally true in our field sites, both primary
and secondary (some places had more diversity than others among those
people with powerful positions), but across the overall landscape, this was
often the case. As the definition of White privilege involves a certain blind-
ness to one's own advantages, we found that, in general, more people of color
were aware of, and focused on, these connections than were White people.
However, we also found that some White people, particularly older White
men, were very clear about how the racism they encountered as children still
exists today, demonstrating their reflexivity, acknowledgment of their own
White privilege, and ability to speak frankly about touchy topics while being
neither defensive nor dismissive.

As a general focus on perceptions of race and racial relations is the necessary backdrop to our more specific exploration of race in the context of local
bioenergy developments, we focus on this first. Then, second, building on
the discussions in chapter 3 about values attributed to forested landscapes,
we will explore the myriad ways in which Black Americans see and value
land generally and forested land specifically. We frame these valuations in a
discussion of cultural landscapes, showing how several values stand out as
having particular importance for Black Americans, especially for landowning families; these include family legacy, a cohesive Black community, and
religious values. Third, we discuss the challenges and opportunities of Black
land ownership. We show how many Black landowners struggle with gaining clear title to family land and the obstacles that it presents for choosing
the best land management strategies for their land, their goals, and their lifestyles. We focus in particular on the concerns associated with heirs' property,
or land passed through generations without will or clear title. Fourth, we elucidate our recent findings on current forest management strategies of Black
landowners, and how their value systems affect land management decisions.
Drawing from our own research and related research by our colleagues, we
demonstrate some of the main differences between Black and White landowners, as well as diversity of management approaches among Black landowners (Goyke et al. 2019; Goyke and Dwivedi 2020). Fifth, we explore the

various ways that information about forests, forest management, and potential new developments in wood markets circulates within the Black community, in addition to the ways that Black Americans are influenced by many of the same factors as the population at large (i.e., social media, nongovernmental organizations, mainstream news outlets) (Hitchner et al. 2019). Sixth, we explore the perceptions of the Black Americans we interviewed of the feasibility of wood-based bioenergy in their communities, their thoughts on the potential impacts of its development on the natural landscapes, and their concerns about who benefits and who pays the costs of bioenergy development in rural, forest-dependent communities. We end with a discussion about how the issue of race is given a peripheral role in the broadly circulating bioenergy imaginaries that we have already presented and how an approach more focused on the nuances of racial dynamics in small, rural, forest-dependent towns can elucidate how narratives about the costs and benefits of bioenergy development are not smoothly translated across scale.

Here we feel the need to call attention again to several factors that undoubtedly affected the results presented here: (1) our status as highly educated White academics (two senior men and one junior woman) representing Georgia's 1862 land grant university (University of Georgia) and the USDA Forest Service, a federal agency; (2) the tenure of Barack Obama as the first Black president (he served as president from 2009 to 2017, and our fieldwork covered both the end of his first term and the beginning of his second); and (3) the timing of our fieldwork as before the Black Lives Matter movement took root and before COVID-19 exposed flagrant disparities in both our health care system and extra vulnerabilities of those workers deemed "essential" (many of them lower income people of color).

We also want to acknowledge our own struggle to accurately portray what we heard and witnessed without perpetuating stereotypes of members of any race. We recorded a number of illustrative quotes that we cannot use because they reveal too much about the speaker; even though we do not identify communities by name here (we draw from interviews all over the South, not just the ones from our primary bioenergy field sites), sometimes there is just no way to adequately protect identities of people and communities. We also had to make hard choices about what to publish when people used racially offensive language, and there are some words we have chosen not to print at all, though we indicate when they were used. Some of the language and sentiments expressed in the quotes from our interviews on Black history and relations in communities are uncomfortable to read at best and may be quite jarring to many. We have left out what we found to be the most offensive quotes, which we did find to be outliers. However, our aim in this book is to capture and examine the many ways that people talk about bioenergy both directly and indirectly, as well as issues linked to it. As explained previously, we do this through a focus on everyday talk, the actual words and phrases that people use, which we believe provide necessary context for understanding the social dynamics within communities and the relationships that various people and communities have with their landscapes and with larger economic and political systems.

RACIAL HISTORIES IN THE LANDSCAPE

Those Here First: Recognition of Native American History in Rural Communities

Prior to the arrival of European explorers and settlers, the southeastern United States was home to dozens of Native American tribes; some of the larger tribes included the Chickasaw, Creek, Cherokee, Choctaw, and Seminoles.[1] They hunted, fished, gathered, and farmed a number of crops such as corn, tobacco, beans, and squash. Many perished directly through violence or indirectly through exposure to diseases, and many have been expelled from their traditional homelands, through forced migrations such as those on the Trail of Tears (enacted through the Indian Removal Act of 1830, signed into law by President Andrew Jackson). The families of some have remained in the southeast or returned there after removal. In some areas, visible markers of the past, such as mound sites, stone monuments, and rock effigies, have been preserved or restored. However, many more sites have been lost.

While Native American culture and history were not central to our research, some people in various states mentioned this history, artifacts they (or others) had found, or family lineages including "Indian blood." In many of our field sites, people talked about finding arrowheads, especially in fields along rivers. One local historian in Columbus, Mississippi, mentioned a large river that was once a political boundary between tribes: "The Tombigbee River divided the Choctaw from the Chickasaw. This area was also contested by the Creeks." He told us about the Choctaws who once lived there, and how they sometimes held higher socioeconomic statuses than White citizens, as evidenced by the expensive china they had acquired, although "by 1820 or 1830 that changed, of course." A woman in Georgia told us that "There were a lot of Indians in the area. In the late 1800s, settlers were still fighting Indians. It's a shame, because they could have learned more from them." She said that many farmers had simply plowed Native American sites over the generations. A man in Georgia said that his "family married into Indians." He said, "I have relatives in Cherokee [North Carolina]. My great-great grandmother was a full-blooded Cherokee." A Black historian in southern Georgia spoke of the Native American history in and around the Okefenokee Swamp.

He said, "There were Native Americans here long before anyone else was. In 1824, soldiers came through Georgia, set up camps, and drove the Native Americans into the swamp. It was like the Trail of Tears. Very unsettling." He spoke at length about the "forgotten" history of Native Americans in the region, and some of the ways that he has been trying to preserve it by consulting historical records to document place names, family histories, and gravesites. He said that some of the history could be read in the current landscape, even in its infrastructure, as:

> You see railroad tracks, utility poles, rivers all in the same area. Railroads and telegraph lines followed Native American trails, which followed rivers. Then they became roads, or roads put near them. All covering up the Native American history. From Waycross to the Gulf, it was like that. One long Native American trail.

He also visited and interviewed Native Americans still living nearby, saying, "People are surprised that we still have tribes here. A lot of them [Native Americans] are very private. They change their names to blend in." He is working with tribal members to educate people about derogatory terms such as "squaw." He spoke about the intentional erasure of Native American history, and even compared Native American history to the history of his own enslaved ancestors, saying, "When it comes to Native American history versus African American history, they have it worse than we do." He laughed a bit when he said, "Illegal immigration began in 1492, you know."

While there are many interpretations of archaeological and paleo-ecological records, increasing evidence suggests that the pre-Columbian landscape of North America, including the South, was an anthropogenic one, having been transformed by its inhabitants by land clearing, farming, and fire (Abrams and Nowacki 2008; Denevan 1992; Hammett 1992). Pre-Columbian land and forest use patterns represented a functional landscape that included farmed areas growing corn, beans, squash, and sunflowers; as well as uncultivated areas that were used for hunting, fishing, and gathering of diverse plant materials for food, medicine, and fuelwood (Goodwin 1977; Hammett 1992). Fire was used to clear forest for farming, manage vegetation around villages and towns to prevent uncontrolled fires, manage forest to remove less useful vegetation and promote more valuable vegetation (e.g., nut and fruit bearing trees), and create favorable conditions for hunting (Goodwin 1977; Hammett 1997; Newfont 2012). After European contact and steep declines of indigenous populations due to introduced diseases, secondary forests grew up on abandoned fields (Delcourt and Delcourt 2004). These regenerated as late successional ecosystems and were erroneously seen by Europeans as wilderness (Delcourt and Delcourt 2004). As European settlement proceeded, forests were cleared, extensively logged, or further ecologically modified. In the South, large-scale land conversions to plantations of crops such as rice, indigo, cotton, and tobacco became prevalent, and these depended on the forced labor of enslaved African people.

"Wrapped Up in Slavery": Black History in the South

A number of people of different races across the South expressed what one White man told us about his town: "The history of the place is wrapped up in slavery." Others, mostly White, skirted around the issue of slavery or never brought it up, because it did not seem relevant, was too difficult to talk about, or because the thought never crossed their minds. But slavery cannot be separated from the history of the South, nor from its ongoing legacies. Both Black and White members of the communities we visited spoke about visible reminders of slavery in the landscape. One White man moved to another town in the South as a child and said, "There was a slave market there; they used it there. They used to sell Negroes there. When I was a kid, we used to play in it . . . Now there's a monument there."

The history of slavery is also evident in the shared family names of Black and White residents of the same towns; it is commonly known that the forebears of some Black families were once enslaved by some White families, only a few generations removed. One White man, who grew up believing that his family had been too poor to own slaves, related this story of his discovery that, in fact, they had. He said there were other people around with his last name,

> and not all White, you know. I always heard that there's no Black [*family name*] because we weren't rich enough to own slaves. But the other day, I was at the [*store name*], and there was this Black lady ringing me up, and you know what her last name was? [*family name*]. I told her: "You and me must be kin." So I know we did used to have 'em. We all knew it. My nephew said to my daddy once that there must be some Black blood in the family, cause, you know, there just is when your family has slaves, and he said, "No, there ain't." But we all know there is. That makes my son real mad to talk about it, but he knows it too.

In other towns, we would mention a family name, and people would sometimes ask: "You mean the White [*family name*] or the Black [*family name*]?" The association to past slavery was too obvious for most people to even elaborate.

Enslaved persons, and later freedmen and sharecroppers, labored on row crop farms, mainly cotton and tobacco, throughout much of the southeastern United States. Several older men told us of their memories of these agricultural systems (we further discuss the experiences and memories of Black Americans in the following sections). A Black man told us, "There used to be tobacco, turpentine, cotton. All that's changed." Other people also spoke of the extensive tobacco farming and the Black farmers, farm laborers, and sharecroppers involved in its cultivation. A White man said, "There used to be a lot of tobacco here. There was an old Black sharecropper who used to pay his rent in tobacco." A middle-aged Black man said that his father and his uncle worked in turpentine, and when he grew up, people around him mostly grew tobacco. He shared a memory of his own labor as a child in the tobacco fields: "In the 7th through 12th grade, that's how you paid for your clothes. You work in the tobacco fields in the summer. That was hard work." Cotton, often known as "King Cotton" in the South, is probably the crop most associated with slavery in popular culture, and it was once ubiquitous across the southern landscape. Cotton farming has decreased, and has become mechanized, requiring less human labor. But in recent memory, it was largely produced using Black labor. One White man said,

> Farming has changed. It's more industrialized and more centralized now. I still remember cotton with mules and Black people. And the John Deere, the 4020 tractor, the old type. You can plow 400 acres in one day now. It used to take 3000 people.

Black Labor in the Forested Landscapes of the South

Enslaved and then free Black people were also intensively involved in the strenuous and dangerous work of the timber trade: felling and debarking trees, transporting logs, and operating lumber mills. In the extensive pine forests (predominantly slash pine and longleaf pine) that stretched from Virginia through the Carolinas, south to Florida and west to Texas and Oklahoma, the production of turpentine and naval stores (tar, pitch, gum, and rosin to make ships watertight) from pine trees also relied on slave labor (Outland 1996); after Emancipation, many formerly enslaved persons continued to work in the timber and turpentine industries under very similar conditions (Johnson and McDaniel 2004). A number of people with whom we spoke, both Black and White, shared knowledge and memories of the pre- and post-Civil War labor of Black men in these forest industries. One older White man whose family established the timber business that he and his children continue to operate told us,

> It took many Black men to cut and transport the wood. It's really dangerous to cut. You have to look at the limbs and the wind, how it changes. It's very dangerous to cut trees. A lot of good men died.

The turpentine and naval stores industries were another predominant bene-factor of Black labor, both during and after times of slavery. Several people mentioned turpentine stills in their hometowns and surrounding towns, which were "built with slave labor" as one man told us. One man in south-ern Georgia (the region which "used to produce fifty percent of the world's turpentine," according to one interviewee) told us that "towns were paid for by dipping tar." A number of enslaved men worked in forests, and after Emancipation, Black labor continued to fuel the turpentine industry, which was notoriously labor-intensive. A Black man told us that "A lot of slaves died of diseases spread by mosquitoes, and just from hard labor." A White man described the local naval stores company after Emancipation, which was run by a wealthy White family that owned thousands of acres (and formerly hundreds of slaves):

> Dozens of Black men would be there each week to get their checks. They all worked for the [*family name*]. They had quarters that they all lived in. You might can see one or two of them. I think someone still lives in one of them.

Another older White man, in his eighties, explained that "at those old mills, the laborers were paid in scrip. That was how they kept them enslaved—they'd have to buy all their goods at the commissary." Another White man told us, "There was a lot of exploitation of minorities, of Black labor, in turpentine . . . and between you and me, there's still a lot of exploitation of Blacks."

It is worth noting that several people in southern Georgia mentioned a place in the Okefenokee Swamp whose name we cannot print here (racial slur beginning with "n," hereafter written as "n******"), known as N****** Island. Several White men and women mentioned it to us using its full name or a variation (some refusing to use the word itself but making sure that we understood what was meant). One White person told us, "There's one island, N****** Island, that's where the slaves lived when they were doing timber harvests and turpentine." Another man, a local Black historian who has conducted many interviews and consulted numerous historical records about islands occupied by enslaved persons and employed free Black people, talked extensively about this place, even laughing about its name:

> Ossie Davis [Black actor from southern Georgia] . . . asked me about N***** Island once. N*****, spelled N-*-*-*-*-r. That's what it's called. Some people call it N***a—that's spelled N-*-*-*-a—Island, to keep up with the times [laughing]. Or called Negro Camp Row. I knew about this place, but I couldn't find it on any maps. Slaves lived there, and Black workers . . . I looked at the maps in town, I talked to a lot of foresters. Mostly I want to look at the

contribution of Blacks to the development of the swamp . . . I want to put it on maps, show it to people so they know. Even when they were still in servitude, they did participate in industry. They lived in these little shanty houses working in the timber industry . . . N***** Island, all the buildings are gone now on the main route. Anyway, you can't see any of the old structures. It was close to the railroad. They would cut the trees or get the turpentine and take it by rail . . . there were rails on the edge of the swamp, not in the swamp. They use carts and wagons to get the trees to the rail cars . . . The swamp . . . for Seminoles and slaves, it was a place to hide. But it was more than that. There was a social life there.

Turpentine is mostly a thing of the past, though there are experiments with new techniques in southern Georgia today. Turpentine antiques are available in some of the shops in small towns, and when visiting one Black-owned antique store, the owner showed us some old tools, tar and turpentine cups that were once nailed to trees, and an old wooden turpentine barrel that still had some turpentine in it. He mentioned an upcoming annual forest-related festival and said that "people come from all over to buy stuff like this."[2]

The Times Are Changing: Rural Racial Turmoil in the 1960s

The South underwent a profound change during the civil rights era of the 1960s, which dismantled discriminatory laws related to segregation with the force of the federal government but instigated a much slower change in everyday behavior and attitudes in the rural South. A few interviewees, both Black and White, talked about the process of integration in their small rural towns in the South and related their own experiences with it. One White man told us:

> Before integration, there was KKK. Once a month, the lights would go out. They'd march the main streets of [*town name*]. I was scared to death. I hated it. I didn't know what it was. A few years later, I saw a cross on fire. Back then, if a colored man even looked at a White girl, they'd whip him. I never went along with that program.

He also described an incident in 1960 or 1961, which he found very scary:

> I was working at the movie theater at the time of integration. One night, I didn't think nothing of it, then these 300 some Blacks come in. They said they didn't want no trouble. Said they just wanted to see a movie, on the floor and not in the balcony. I was in the ticket box, alone. I was so scared, you know. I said, "I need to call [*name*]." He was the owner, he had 16 theaters. So they said I

could call him, and I had to walk through them to get to the telephone. I was so scared. They just parted to let me through, and I walked through them to the telephone in the other room. I closed the door and I called [*name*], and I told him what was going on, and I said, "What do I do?" He said to just wait it out and stay on the phone and don't do nothing. So we talked for a long time, and when I looked back out, the lobby was clear. They all left, and they never did try that again. I tell you, I never been so scared . . . . I got out of the Army in 1962. When I came back to the theater, they'd closed the balcony, even let 'em downstairs. It was all separate before then. They had the same stuff upstairs as downstairs, even had a colored cashier upstairs.

The parallel in this story to the intimidation tactics used to terrorize Black people is difficult to miss, as is the reflection of the nonviolent approach to civil rights reforms adopted by many Black Americans in the 1960s. People also spoke of the lingering effects of slavery and Jim Crow era segregation in terms of current racial inequalities and prejudices. One Black man told us, "We all have some prejudice. I never hung around with White people much. My siblings went to [*school name*]—we were the second family there when it was integrated. I was in 6th grade when they integrated it." People, both Black and White, also mentioned private schools in towns and cities as a continuation of segregation for the explicit purpose of the perpetuation of White privilege.

A number of people in all states in which we worked specifically mentioned the changes in racial dynamics over the past several generations; some said things had improved, while others said they had not. One older White man told us that he "surprised himself" by befriending a Black man within the last few years. They met while the Black man was doing yard work in his neighborhood, and

one day he and I got to talking. And then we got to talking more and more, and now we talk all the time. And you know what? I don't see him as a Black man anymore, I just see him as my friend. And I told him that too, I told him, "You know what, I don't see you as a Black man, I see you as my friend." And I know people talk about it, you know, me talking to a Black man so much, but I don't care. I told him that too. People are always gonna talk, but I don't care if they say to each other they seen [*his own name*] talking to a Black man like he was his friend. He is my friend.

He then mentioned that his granddaughter, also White, appeared to not notice skin color in her choice of friends, much to the dismay of her father (the speaker's son):

things used to be all separated. Blacks and Whites just didn't have nothing to do with each other. Different schools, different churches, just didn't interact at all.

> Then everything got integrated, and now it's no big deal to kids to be with other
> kinds of kids. I took my granddaughter to a basketball game at her school, and
> she wanted to sit with her friends, and I looked over and she was sitting with a
> little Black girl and they was just chattering along like they didn't even know
> they was different. Didn't even see it. Couldn't see it. My son, he didn't like that
> too much, but that's just the way things are now . . . kids these days are just used
> to that. They're used to being around 'em, and it's no big deal.

People told us other stories about how race relations in these rural towns
had improved, though some seemed to imply that instances of kindness and
cooperation were more isolated than widespread. Nonetheless, they were
imprinted on the minds of people, both Black and White, as encouraging
signs. One Black historian discussed his efforts to document and preserve
Black graveyards and gravesites, many of which have been lost to history.
But he told a story of one Black cemetery that is now in a White community:
"the Whites are taking care of it, preserving the history. They show a lot
of respect, and that's a good thing." Another Black community member, a
woman, said that change was coming, albeit slowly: "They still call it 'The
Dirty South.' Maybe it won't always be. I'll not be around for it, but I've
lived in it. Things are getting better, but slowly." In one town, a Black community member commented on the lack of diversity of local decision-making
bodies but suggested that things were soon likely to change. Emphasizing
acknowledgment of the past, with an eye on the future, he noted that these
coming changes could allow for more opportunities for the Black community:

> There's one family here that holds authority. But the hearse wheels are rolling.
> The younger generation isn't so involved . . . . Let's get away from this out of
> the box structure. We're in the 2000s now, things change. Let's not play the race
> card. We've not yet got our full share, but we're further along than we were.
> Let's work together to make more advances.

Several people also talked about demographic changes among local county
and city decision-making bodies such as county commissioners, city councils,
development authorities, and departments of health and education to include
more Black Americans. We were present on more than one occasion when a
new Black member was voted into one of these groups as a direct response to
calls by Black group members or community members to address a current
lack of diversity. We cannot discuss the details of these instances, but some
people, mainly Black, saw this as a positive sign, while some others, mostly
White, expressed concern that, as one person stated, "they're taking over."
   Fears about minorities "taking over" run so deep as to almost seem cliché, but they demonstrate the ways that while civil rights have made great

advances in some ways, racism remains a real and ever-present part of daily life in the South (not unique to the South of course). One Black man said that while the racism is not as overt as it once was (i.e., burning crosses and lynchings), "there's still prejudice here. Everyone's got some prejudice. So it's still here, but people don't act it out like they used to." One Black woman simply said, "It's not going to get better. It's the South . . . The South be the South. The South will always be the South."

"I'm Not Racist, But . . ."

In talking to people about their communities, the topics of race and perceptions of racism were never far from the surface. We found a number of White people, some already quoted here, that spoke reflexively of the racism in which their childhoods were steeped and their own efforts to reprogram their own attitudes toward other races. This explicit acknowledgment of White privilege was particularly evident in some of the older White men with whom we spoke. Sometimes, it seemed that while people attempted to address their own racism, they were still caught in a mindset that accepted the status quo. For example, an older White row crop and catfish farmer told us:

> This place has changed. It used to be a good place to live. Racism on all sides destroyed this place. It's not that I blame them. If I was Black then, I'd want things to change too.

Though he acknowledged that the demand for civil rights was justified, he also lived under the assumption that "it used to be a good place" when boundaries between races were more secure. Black people and other minorities did not universally agree that "it used to be a good place." As we explore in an upcoming section, many Black Americans did mention nostalgia for the more cohesive Black communities of the past, but that sentiment does not translate to agreement that a segregated society based on racial discrimination, intimidation, and violence was "a good place."

In any case, we found numerous similar instances of White people saying racist things, then explaining to us in detail examples of why they do not consider themselves to be racist. This "I'm not racist, but . . ." mindset was evident in all states in which we did research. One White man, portraying himself as not just tolerant but politically progressive, told us he had been part of a series underground (literally—meetings were held in basements and bunkers) meetings in the 1960s with Black Panthers, gays, and feminists:

> You know who got Obama elected? Gay intellectuals. In the 1960s, I got beat up standing up for my gay friends. I also burned 200 brassieres, standing up for

your [looked at Hitchner] rights. You know the Human Rights Campaign[3]? It was started in 1983 as an umbrella for gay and lesbian organizations, GLAAD,[4] etc. They're the ones that got Obama elected . . . Thirty years ago, they [Black and gay intellectuals] had a plan to become teachers, TV producers, professors, to infiltrate the schools, the universities, the media, and then take over. And that's exactly what they did . . . . Obama was created by the press. He was nothing before.

Denying Barack Obama's intelligence, capability, and vision, saying he was "nothing before" he "was created" stripped Obama of both agency and humanity. Other White people also displayed intense dislike of President Obama, hardly bothering to hide the racial overtones of their criticism. In one community in which we worked, an organization that had been previously federally funded had recently shifted to become a nonprofit. One of the main benefits of this change, according to one White woman, was that: "We can take down all the civil rights posters and Obama's picture. We put up Reagan's picture. We found it in a desk drawer."

Another story we heard from a White man illustrates White privilege disguised as anti-racism. He told a story about a sheriff in a nearby town who was indicted for police brutality. The man said:

He'd been arresting Black people and taking them to the jail and beating them this way up and this way down with his stick, and everyone knew it, but no one could do anything to stop it. He was terrorizing people, and it ain't right. You can't do that to people, I don't care who they are. One day, in front of the [*restaurant name*], he pulled over a car driven by a Black man. He pulled the man out of the car and slammed his head down on the hood of the car. Another Black man was in the parking lot of the [*restaurant name*], and he crossed the road and said, "You let that man go." The sheriff looked at him and said, "I don't know who you think you are, but I'm the sheriff, and this here's my jurisdiction. So you get on back." The man pulled out a badge and said, "Well, I'm a federal marshal, and I've had my eye on you. You let that man go, and I'll see you in [*name of closest big city*]."

Though the story had the ring of being told many times, with embellishments here and there, there is likely some truth to it. However, it is told as a funny story, with the arrest of a White sheriff by a Black man who outranked him as the punchline instead of genuine concern for the victim or the systemic racism that allows these events to continue to occur. Another community member, a White woman who had moved to this town a few years ago from out of state, said, in a pleasant tone, that she loved it here and that "The Black people here are so nice, so polite."

Meanwhile, other White people tried hard to convince themselves (and us) that they were not racist, even while using the most egregious racial slurs. For example, when one of us mentioned meeting with a group composed of Black Americans, one White man said: "That's a bunch of n*****s, ain't it?" He proceeded to differentiate between African Americans ("honest, upright citizens"), Black people ("regular people, they can't help who they are"), and n*****s ("trash"). Then he gave several examples of why he's not a racist. He talked about the respect he had for his first Black teacher and his financial donation to a Black politician. He said, "You know why I did that? Because he's the most qualified. But I had people calling me up saying, 'What the hell are you doing, giving money to a n*****?' But I didn't care. He was the best man for the job." In this man's mind, he is not, in fact, a racist. This example also underscores the reality that these racial epithets are alive and well in the South, not at all relics of the past; children grow up hearing them, and as adults continue to use them.

However, this concern about being seen as racist spills into (and feeds) racial stereotypes perpetuated by prominent individuals, social media, fringe media, and mainstream media such as news outlets, magazines, television shows, and movies. As many shows and movies are shot in the South, there is always concern and speculation about how racial dynamics are portrayed on set. One documentary was set in a town in the South that we visited, and one White community member said he worried about how the town would be portrayed. Using particularly colorful and offensive language, he said:

> Well, they didn't tell us they were doing it. We heard about it and man, we were scared. Our fear was how they were going to portray [*town name*]. Go out in the ghetto and film a bunch of barefoot Black kids and people out on their porches smoking crack rocks and shit, you know the stereotype. They said later, "We should have let you know." But I got no problem with the story. It was fair, right down the middle.

Other people, both Black and White, addressed other stereotypes of rural Black Americans, mainly involving early pregnancies, addiction to drugs, and reliance on welfare. One White woman, speaking of Black Americans, told us, "The majority in the county don't pay any taxes, don't have a father figure at home. There's no real incentive to improve themselves or their children's lives." Her (White) husband added, "Biofuels coming here would have a minimal effect on the way society is in this county. It's a very poor county. It's been imbued for three generations for a minority girl . . . to get pregnant and get food stamps." A Black man stated, "Now we have a Black president. That proves to people that Blacks and Whites are the same. But there's this perception that there are more Blacks on welfare than Whites. There's lots

of propaganda." One White woman, referring to minorities receiving government assistance, said at a public meeting:[5]

> What were you talking about when you said people would starve if we use food
> for fuel? Did you mean people in the U.S. or elsewhere in the world? Because
> no one will starve in the U.S. They get food stamps from the government and
> buy better stuff than I do because it is not their money. My brother has a grocery
> store, and he says they don't care how expensive it is. They can even get food
> stamps for pet food. I love animals, but I pay for my dog's food.

However, a few White people (in addition to multiple Black people), noted
the irony of White people stereotyping Black Americans as lazy welfare
recipients. One White man said:

> People think, the pork coming to me is good. All the other pork is bad. I don't
> get why people vote Republican when it's not in their best interest. They should
> all vote Democrat and get the money [laughs]. People bitch about welfare but
> get lots of farm subsidies. Don't they know that welfare gets less than 1% of
> the national budget?

This echoes the views on subsidies discussed in chapter 3, only here with
a clearer connection to racial stereotypes; essentially a lot of people mean
that White people getting money from the government is okay, while Black
people getting it is not.

Acknowledging Limitations and Providing Context for the Stereotypes

We also encountered criticism for lower income, in some cases, welfare-
dependent, Black Americans by other Black Americans. One Black man told
us the root of the issue is a combination of laziness and hopelessness among
minorities, especially Black men; he said, "Uneducated Black males don't
desire more than what they have. They have opportunities to be educated,
and they choose not to be. They want employment, but they don't want to
meet the requirements." However, this criticism was usually couched in a
deeper understanding of the specific challenges faced by Black Americans in
rural southern communities, through their own lived experiences. One Black
community leader said, "Even after slavery, many people never left. That was
home. But the system didn't change much. There are generations on welfare,
it's always been like that." An older Black man elaborated on the changes
in the community after most economic opportunities for Black community
members dried up:

[*County name*] County was all turpentine and farm. The community can't function now. I guess all small communities have a hard time. Here there's a lot of unemployment and drugs. Shops come and go like birthdays . . . It's bad here. I've been here 40 years. It was different when I came here. Now the place overflows with drugs, like in every small community.

The loss of employment opportunities for all young adults, and, in particular, for young Black men, has taken a drastic toll on the community, through high rates of outmigration to cities; problems with drugs and violence, often leading to incarceration; and reliance on government assistance (figure 5.1). Another Black community member said, "This county is hurt more than most counties. Most minority families . . . have to travel to get jobs. Long ago, they did turpentine. Now that's gone." This sense of hopelessness was expressed by a number of people. One Black man said, "I was born and raised here. I see the potential. It hurts that there's no new growth. Nothing has changed in my lifetime . . . . People have this negative attitude that nothing will work. It hurts me to my heart."

Drugs, and drug testing as a requirement for employment, are perennial problems in the rural South, as is the disproportionate link between drugs and incarceration among Black Americans. One Black man told us about the high concentration of prisons in his region of the state in which he lived. He said, speaking primarily of young Black men in rural areas:

People knock on the door of jails, because of the economy, culture, drugs. 95% of the people in jail are there for drugs. There are drugs in every small town. I've seen a 25-year-old with not a tooth in his head because of meth. Lots of drugs

Figure 5.1   Soperton, Georgia, 2011. *Source*: Credit: Sarah Hitchner.

... prescriptions, meth, good old crack cocaine. They need actual rehabilitation, or they'll just go back to the same situation. After their first failure, they get one charge, and they can't get another job.

While we were in one town, a man employed by the town had been fired for either failing or refusing to take a drug test, and the issue triggered a series of meetings of Black citizens, public hearings, and vocal disagreements at government meetings open to the public. The root of the issue, according to the Black community members who discussed the case privately with us, was that the man was Black and qualified for his job, while the man who replaced him was White and not qualified (the drug test itself was also considered a set-up, an excuse to replace this worker). We cannot reveal details here, but it became obvious that the case clearly demonstrated the fissures in the community along race lines; everyone seemed to pick a side with those whose skin color matched their own. The case also starkly demonstrated several barriers to employment at most any new industry, including those companies that come to rural towns: widespread issues with drugs and discrimination based on racial stereotypes of associations of minorities and drugs.[6] These issues are interrelated, and small towns rarely have the adequate resources to assist residents with drug rehabilitation programs, or the compassion necessary to try to find funding for them.

Educational Deficiencies in the Rural South

Another major issue affecting local employment, aside from lack of opportunities and drug addiction is what many community members, of all races, consider to be substandard education available to children and young adults in their communities. One Black community member said, "There is significant poverty here. It's not the best school system. You can graduate in [*county name*] and still not know what a fraction is and not be able to read well or how to use computers." One young Black man, a Muslim, talked about the lack of educational opportunities at school, at home, and in the community; he was pushing local parents, school administrators, and elected officials to do more for children's education:

I'm going to talk to the mayor [about problems with the current educational system in the county] and see what can be did. There's no YMCA, no laboratory. The only thing they have is the library. We need to have people to help. Science rules the world. The problem with the system is that there's no challenge. [*town name*] can't compete with bigger places. They don't push education enough here. I tell parents it starts in the home. Your parents pushed you to be a researcher. Kids think small. Teach them to dream big and think big. You can't waste your money on foolishness. Stop letting them watch TV, Facebook. Need to be competitive with other countries. Don't allow people to waste time . . . Science is fun, math is fun, reading is fun. Kids are failing because they're not doing anything at home . . . . I went to Mecca in 2005, saw over 65 million people there. I came back to America, I wanted to shake my head and cry. People don't try and they don't pay attention . . . In rural towns, there's nothing available. What's hard about giving kids something to do. Half the kids have never been to the forest.

It is worth noting the attention this speaker draws to the lack of interaction with nature among many children, even in rural, forest-dependent communities. However, the larger issue he is referring to is educational failure, seen as a societal failure by many Black residents; this has important implications for preparing local community members for employment in emerging bioenergy facilities.

EXPECTATIONS OF LOCAL BIOENERGY DEVELOPMENT

When we asked community members and leaders about the racial dynamics involved with local bioenergy development in their hometowns, we got a lot of answers. They generally fall into the following main categories: (1) how bioenergy development could potentially benefit minorities, and how they have, or have not witnessed or experienced these benefits, and (2) what risks or costs does bioenergy development impose disproportionally on minorities? The answers to these questions quickly become encased in issues of scale: most people were thinking in terms of the immediate costs and benefits for themselves and people within their own communities. Many other people were thinking in broader terms, about how increased bioenergy production could contribute to those larger narratives we have already discussed, such as national energy security and independence, greater ecological sustainability, and bioenergy as a mechanism for climate change mitigation. Many people of color, of course, were thinking about the issue on multiple scales simultaneously; for many of them it was painfully obvious that their specific circumstances, and the challenges they face as Black (or other minority) individuals, precludes their active participation in the bioenergy imaginary that is considered a "win-win-win" because it addresses that "triple bottom line" of enhanced environmental, economic, and social benefits.

On a smaller, more immediate scale, many Black Americans with whom we spoke noted the hopes they had for bioenergy bringing positive developments for their families and communities. For many minorities, the opportunity for employment was seen as the main benefit. These opportunities, they felt, would have a domino effect on some of the other main challenges they faced; more jobs would inspire more young people to do well in school and attend trade school or college. It might also lead to local governments placing a higher emphasis on increasing educational standards and opportunities, and other types of opportunities for children with known benefits, such as recreational facilities, sports teams, field trips, mentoring programs, and so on.

A number of community members, community leaders, foresters, and bioenergy industry representatives also saw the potential for increased revenue to minority landowners, either farmers or forest landowners, through sales of agricultural products (including trees) to bioenergy facilities as feedstocks. As noted in chapter 3, the general scenario for private landowners regardless of race was disappointment with the lack of increased markets for agricultural products and trees and tree products as feedstocks for bioenergy plants. Private landowners across the region noted the failure of the facilities to take waste wood (leftover slash from harvests) or even thinnings, often commenting on how owners of smaller acreages could not compete with the large industrial and institutional owners of forest land such as Weyerhauser, Rayonier, Plum Creek, and Hancock Timber Resource Group, each managing hundreds of thousands of acres of pine trees in the southeast (Zhang et al. 2012). Bioenergy facilities often purchased in quantity from them; in some cases, this filled a void left by now-closed paper and pulp mills, while in other places, it put bioenergy in direct competition with them. Private landowners were thus already at a disadvantage when it came to selling feedstock to the bioenergy facilities. Minority landowners, especially Black landowners, face their own additional unique challenges, which are often rooted in slavery, segregation, prejudice within government agencies, and a host of other manifestations of inequality and discrimination.

In the next few sections, we focus our attention on Black landowners: what land ownership means to them and what cultural values they place in the landscape; specific challenges faced by Black landowners; analysis of the factors affecting land management decisions by Black landowners; and exploration of ways that information about forestry and new forest opportunities flows among Black landowners. These are vital issues that underlie perceptions of Black community members about whether bioenergy development will ultimately help, or hurt, them in particular, and the larger Black community in general.

BLACK CULTURAL LANDSCAPES AND FORESTRY

We begin this discussion with a brief synopsis of work we have done with minority landowners in states across the South that focus on Black land ownership and the values that Black landowners place on land and land own-ership. Our studies show that for many Black landowners, the connection to the landscape is alive and strong. Family land left to them by their predeces-sors, who obtained the land during times of slavery, Reconstruction, or intense Jim Crow racism, and land that often literally contains the bones of their ancestors, has more than sentimental value; ties to the land are deeply embed-ded in their identity as individuals, families, and communities (Merem 2006; Dyer and Bailey 2008; Dyer et al. 2009). Many Black families feel that their ancestors left the land to all the heirs to be held jointly by them in perpetuity as "homeplaces" that "were established originally as a challenge to racism—to repossess a sense of self, heritage, and family security" (Goluboff 2011, 390).

What we encountered during our research with Black landowners was a depth of feeling of land as a repository for childhood memories, legacies of the ancestors, and nostalgia for strong and cohesive Black communities that struggled, and often thrived, within the wider community in which they faced a number of threats and obstacles. Sometimes this connection to the land goes back to a certain beginning, as one man explained when talking about his return to family land in Georgia:

> About 1826, a slave ship landed in lowcountry South Carolina. On that ship was a mother, father, and young daughter. They were sold. Man sold them to differ-ent plantations. The mother and daughter were sold to a plantation in [*county name*]—that was my great-great-great-great-great-grandmother. That's why I came back here. That's the point of origin.

More commonly, people told stories of their childhood memories growing up on or visiting family land. Many described subsistence and row crop farm-ing and working in the fields with other family members. They described the overall farm landscape, including crops, trees, pastures, and fields. One person stated: "We raised hogs, children. Grew cotton, corn, peanuts. Milked cows, churned butter, had pecans, pears. It was a self-sustaining property. There was a grits mill and a syrup mill." Many people spoke of the coopera-tion and interdependence between family members that the family farming lifestyle fostered. One person said:

> When we dug potatoes, it was really fun. My aunt would come over. My dad would drive the tractor. We'd finish ours, then dig up my great-aunt's, then my aunt's. It was a big family thing. People shared what they got. It went beyond the immediate family, to family members in Maryland and other places.

Others spoke about their sentimental attachment to the land as a result of associating it with family. One interviewee, who visited her grandparents' farm often as a child, stated:

I remember going on the back porch—my grandmother would do a lot of things on the back porch. We'd snap the string beans and the peas and harvest the vegetables. They didn't have running water—they had a pump. I was taught to get water. Being a city girl, I didn't know about that. That water was cold and crystal clear. There's no taste like that water . . . I'm interested in preserving the history of the land.

These experiences on family land led to a sense of community, because many properties were in Black communities or enclaves. Several people spoke directly about the emotional and cultural benefits of living in areas surrounded by other Black Americans and the sense of security this brought. One person in South Carolina said:

So there are memories of security and safety, family, that nothing can bother you. That was in the 60s and 70s. There was still segregation. Society was segregated, but there was a sense of security. The area was all family.

Another person said: "It felt good back then. Everyone in the community knew everybody. If one person didn't have, someone else did. It's not like that now." Many interviewees spoke about the sense of community they felt as children, whether or not they grew up on family land. A pastor who spent summers on family land in Georgia said:

We would play as kids would play. But we did learn how to drive early. I think I was 8 years old. That pick up, he would tell us to deliver this to Miss so and so, and we would take what was needed. I never saw him transact any cash. It was always barter. For the most part, if someone was in need, they might come by, and automatically we might trade eggs for meat or corn for peas or something like that. It was a lovely experience for me. Growing up in the city, all we saw was concrete. We'd go down there, and we could walk for days. They didn't too much worry about where you were, because there was nowhere to go. And they knew you weren't going to be out too late because you knew it was going to be so dark.

Many Black landowners with whom we spoke also spoke about the religious values they placed in the land and in land ownership; many saw land as a gift from God and sustainable management of that gift as their responsibility to be good stewards.

Black families today strive to simultaneously hold onto the past and go into the future, to value land for its role in their family and community heritage while simultaneously preparing a future generation to take over the responsibility of keeping their legacy alive while allowing them freedom to choose their own paths. With the decline in family farming, employment in professions and the service sector, and migration to urban areas, trees and forest management have an increasingly important role in providing the economic returns that support family land.

We have also looked extensively into the specific challenges that Black landowners face in managing their forest land, including insecure land tenure, historical discrimination in government land management programs, and unequal access to information and capital. Black rural landholdings in the South declined precipitously—by over 90%—over the twentieth century due to a number of factors, including outmigration, voluntary sales, foreclosures, poor access to capital and credit, lack of access to farm and conservation programs, illegal takings or land theft, purposeful trickery and withholding of legal information, actual or threatened violence, and various forms of individual and institutional racism and discrimination (Mitchell 2005; Daniel 2013; Dyer and Bailey 2008; Gilbert, Sharp, and Felin 2002; Zabawa 1991; Zabawa, Siaway, and Baharanyi 1990). The number of Black farmers has declined more significantly than the amount of Black-owned farmland (Gilbert et al. 2002), and unfarmed land has often grown up in unmanaged forest that provides owners only limited economic returns (Schelhas et al. 2017). Due to discrimination and lack of access to and trust in the legal system over time, much of this land was passed on to heirs without wills, often for several generations, and is owned as heirs' property, a form of tenancy in common where numerous descendants own a fractional interest in an undivided tract of land. For many Black landowning families, economic returns are not the primary goal of land ownership, especially in cases where land has been passed down through generations and is currently shared by multiple family members; however, returns from active forest management can allow landowners the opportunity to keep land with sentimental value in the family (Schelhas and Hitchner 2020; Schelhas, Hitchner, and McGregor 2019; Goyke et al. 2019; Hitchner et al. 2019; Schelhas, Hitchner, and Dwivedi 2018; Hitchner, Schelhas, and Johnson Gaither 2017; Schelhas et al. 2017). For many Black landowners, economically productive forestry practices are a means to another end, not the end goal.

Owners of heirs' property face a number of hurdles in actively managing their forests, both due to the difficulties of co-owning land with multiple family members and the insecure status of heirs' property (Bailey et al. 2019). Upfront investment in tree planting and timber management for family forest owners typically are facilitated by cost-sharing from government conservation programs, which have a history of discrimination in serving Black landowners and have, until recently, seldom been available to heirs' property owners (Daniel 2013; Schelhas 2002; Schelhas et al. 2017). Black landowners have also long had poor relations with forestry providers, often characterized by exploitation, limited access to information, and difficulty in getting services performed on smaller landholdings (Schelhas et al. 2018). As a result, reforestation often takes place through natural regeneration, which produces lower quality timber over longer time periods and thus lower economic returns. Because obtaining the consent and signatures of multiple owners of heirs' property is often a significant challenge, making legal timber sales difficult, harvests are often opportunistic and timber sold at a substantial discount, adding up to a cycle of low investment and low returns (Schelhas et al. 2018). Again, these obstacles to obtaining revenue from their forests can make it difficult for Black landowners to keep the land in the family.

As a result of this history and these patterns, Black Americans own less land and have less economically productive forests. This, in turn, makes it much more difficult for them to benefit economically from emerging bioenergy markets. Programs to address the issues faced by Black forest owners are mostly only a few decades old and limited in geographic scope, although the multi-state Sustainable Forestry and African American Land Retention Program (SFLR) has begun a process to both address heirs' property and facilitate forest management (Schelhas et al. 2018). Black landowners around our primary bioenergy research sites were not served by any such programs and faced numerous hurdles in forestry and sale of timber, leading to feelings that White landowners with many acres in forests were both promoting and benefiting from bioenergy development while they were not. On the other hand, a pellet plant in northeastern North Carolina is located near the SFLR site in that state, and it has provided financial support to the SFLR and represents a potential market for harvests of woody biomass from low-value timber stands. These outlets for wood from forests that have not been actively managed in the past would be a boon for Black landowners that have historically faced challenges in producing sawtimber and other high-value forest products.

There is evidence that Black landowners place a very high value on land ownership as a family legacy and as tangible representation of the larger Black struggle to obtain and retain land on the path to economic independence and freedom from oppression (Goyke and Dwivedi 2021; Schelhas et al. 2012, 2017). Like other forest owners, Black landowners are not a homogenous group, but rather have been found to have diverse views on forest management (Goyke et al. 2019). A typology of Black landowners in Georgia suggests that while some focus on short-term returns, others either seek to manage their timber or recreational areas for long-term legacy and amenity values (Goyke et al. 2019). One notable difference among Black landowners is the existence of a group of "Indecisive Owners" unsure of their objectives and lacking the confidence to choose among options due to lack of experience, trusted technical assistance, and financial assistance (Goyke et al.

2019). The process of rectifying this situation and increasing participation in forestry through programs like the SFLR hinges on a process of community engagement that builds connections in a way that spreads gradually through the community and fosters peer-to-peer learning among landowners while also engaging service providers (Hitchner et al. 2019; Schelhas and Hitchner 2020).

PERCEPTIONS OF BIOENERGY, ROOTED IN EXPERIENCE

The "Good Old Boy Network"; Or, Why I Didn't Get Hired

As in many rural areas in the southern United States, there are ongoing racial inequalities, and minority hiring issues were an ongoing issue in racially divided towns in which bioenergy facilities were located. In various locations, when bioenergy facilities came to small towns, they promised to hire local people. In a few cases, they did hire local minorities. One White landowner in Mississippi said that he thought he might benefit as KiOR was another new potential outlet for his wood; he also thought that "It'll help Columbus with employment. This Black man came in here yesterday and said he got a job at KiOR. He was very excited about them." A Black community member in Columbus noted that "KiOR has had some impact. People were employed to build the plant . . . . Overall, they are hiring . . . . A few people have gotten jobs there." This same speaker said:

> This kind of company, to make wood chips into fuel, now that's something unique. Are they gonna make it? . . . Raising the quality of life raises the community . . . It's like giving a baby a steak—they can't eat it. Once an idea is taught and explained in school . . . going green, green initiatives . . . it will be well-received. People look at solar panels and say, how does it benefit me? KiOR is part of that education process.

He felt, then, that KiOR can play a positive role in the community not just by providing jobs, but also by serving as a source of information about renewable energy.

However, as noted in chapter 3, the jobs did not materialize at the facilities as many local community members expected. In particular, many minorities did not see many job opportunities for them, as a result of racial discrimination, lack of educational and professional qualifications, drug test requirements, or past criminal records. One young Black man said, "My brother was incarcerated. He just got out of jail. He has no degree. What will he do?" His options were very limited indeed in that town; he would likely never receive a job offer from a bioenergy facility. In some cases, a lack of local hiring was the result of there not being as many positions available as the company had anticipated. In others, people were brought in from outside the community; one reason is rooted in the "good old boy network" in which White managers hired their own relatives and friends, or at least people that looked like them. While some Black community members expressed disappointment that more jobs for them did not materialize, some were not at all surprised. This led some to strive to initiate change and others to remain resigned that life in the towns would likely not improve for minorities. Some felt that they would continue to be shut out of jobs, even ones for which they were qualified.

A few Black interviewees mentioned how at least some of the opposition to bioenergy facilities was coming from White citizens who were concerned about the aesthetics instead of as an opportunity to pull the city or county out of poverty and employ local people. One Black community leader told us:

> What we need is a few good funerals . . . this is still the Old South. There's still some of that, but there's some that's new and different. Some people see a construction dumpster, and they hate it because it's ugly. But I see it as a sign of progress. . . . There's still the good old boy network.

As noted earlier, another Black community leader commented on how generational turnover will initiate openings for more Black leadership. However, as stated earlier, we did witness the appointment of several Black community leaders as a result of community pressure, so death was not the only agent of change.

A few Black community members felt that in some cases, bioenergy companies deliberately misled the community leaders by promising to employ local people, often minorities. They often implied that the bioenergy companies worked hand in hand with the people in positions of power in the town and city in which the facilities were located; these local government officials were the ones that lured the bioenergy companies to their towns, and in turn, the companies let the officials direct most of the local hiring. When referring to discriminatory hiring at the facilities, they used words like "hoodwinked" and "scammed" to emphasize malintent on the part of both the companies and the people in power, who were mostly White. One Black community member explained the context into which these new bioenergy facilities were placed: "There is still a Black and White issue here. It's almost servitude, two notches removed, depending on who they work for." Several Black community members we interviewed commented on the lack of diversity in the hiring processes at bioenergy facilities, and one explained at length what happened in his town:

There's a lot of false advertising for jobs. Good people don't have a fair shot at the jobs. . . . The county ran ads [for jobs]. It was false advertising. They put people in positions that don't even qualify. I can't even get a phone call from them. What's going on in this job market, it's not right. And the [*bioenergy facility*] plant, they did a lot of picking and choosing. One [Black] guy didn't get hired because he was too highly qualified. But the job don't require a high school diploma or a GED. But it ain't advertised that way. So that discourages others [Black Americans] from even applying. . . . It's all White and Black. They won't talk about it in public . . . I want you to know while I'm living . . . they're afraid to print the truth, do the research, on this part of it. And they wonder why they can't get qualified people. They just put up this invisible sign: "No need to apply." Got to overcome that. [*Bioenergy facility*], they didn't know how severe the community is, the imbalance on the board, how it's all on one side of town. So how to balance, how to share with the community? It never occurred to them. And there's still no change, see?

One politically engaged Black community member, recognizing that non-locals were often brought in for jobs created by industrial development that was targeted for poor and minority communities, pleaded for the companies to "tell us how many jobs and what are the job classifications and competencies needed so that we can prepare [*county name*] people for the jobs and contracting opportunities."

In some of the rural communities we visited, concerted efforts were being made to improve educational opportunities for children and adults, including more options for vocational training at community colleges to prepare people for the types of industry, including bioenergy or other renewables, that may locate in their hometowns. The first step, several people explained, is demonstrating the importance of graduating from high school and then pursuing higher education. One Black community leader told us:

I want to see people get the desire to get credentials. People have nothing to offer. I need to think . . . how to educate people in the community about the importance of a GED, Associates, a trade. Have something to offer, instead of being on welfare.

But for employment of local community members at bioenergy facilities to become a reality, many people told us, there need to be programs in place for these educational opportunities and jobs requiring them—and companies willing to hire local people in the first place.

Some Black community members in the towns with bioenergy facilities felt that these companies may have had good intentions—with real plans to hire local people—but were ignorant of the reality on the ground. One Black community leader in Alabama stated:

I assume someone in [*bioenergy facility*] has done research on socioecological factors in the Black Belt and [*county name*]. But I can't swear that's happened. Do you understand where you're coming? Most are from the Midwest, and I'm not sure that they understand. This county is 80% Black. If they say they'll hire local people, will the majority of their workforce be African American? I know the top management will come from them, but will they have Black foremen and forewomen? Will they be willing to work to hire and promote local people? They'll draw from [*names of four counties*]. Will many of the employees be Black and have the opportunity to move up?

Several other Black community members in different states emphasized the jobs that Black Americans had gotten at the bioenergy facilities were low paying with no room for advancement, such as janitorial work. Meanwhile, they saw White locals receive jobs that included additional training that would allow them to learn new skills and make more money. One Black man explained what he had witnessed regarding employment in the county, which will affect hiring at the local bioenergy facility:

It's gonna cause the plant problems, if this discrimination's not dealt with . . . These things in the community, they'll have bearing on the plant. People think they don't have a fair chance, they don't bother applying. This one's working the motor grader, it's a high paying job. Then he goes to superintendent. All the applicants that are Black have diplomas. This one White applicant doesn't have a diploma, and you know who they hired? That's what will affect how the plant operates.

Ongoing, and very obvious, racism will continue to affect hiring patterns in bioenergy facilities in small southern towns. In several cases that were explained to us, the companies did not end up hiring the people that were the most qualified; this, as several Black community leaders pointed out to us, will ultimately affect not only their bottom line, but their very survival. Companies, even if they intend to hire fairly, are often caught in the cross-hairs of local politics and historical racial dynamics.

What Bioenergy Company? How Lack of Information Thwarts Participation

Many Black community members, even community leaders, with whom we spoke felt that they just didn't know much about the facilities located in their hometowns because that information had been withheld from them. One Black community leader told us, speaking of a facility that had closed:

People aren't sure why it closed. Maybe their questions are not answered. People in my neighborhood are out of the loop. They maybe read the papers. Their questions are unanswered, but there are speculations. Maybe they blame the government . . . The council members and board members make decisions. My neighborhood doesn't have any communication with them. If you don't know them, you don't get that information. You have to speculate.

Another Black community leader felt that lack of enthusiasm about bioenergy was caused by lack of information and lack of participation in its planning. He made the following suggestion:

We could have a big workshop, a workshop to introduce the bioenergy plant. Do some visioning, where everyone has a voice. That's the problem with [*town name*]. We need to get momentum, to diversify. How to get the Blacks in my neighborhood excited about biomass? There needs to be momentum in the Black community.

A number of other Black community members also told us they knew little or nothing about bioenergy or had not thought about it much. Given the severe challenges facing their communities, combined with low rates of higher education and limited access to local decision-making processes, this was not altogether surprising. However, quite a few offered broad support of renewable energy in many different forms (i.e., solar and wind), a few specifically mentioning hope that President Obama would steer the country away from fossil fuels. One young Black man said, "I'm in favor of bioenergy. Remember, energy is never created or destroyed. There are a lot of resources here in the U.S. We need to tap into our land right here. President Obama need to look into that heavy."

Expectations versus Realities for Black Landowners

As with landowners in general, some Black landowners envisioned themselves sending wood products (either pulp or slash leftover after harvests, not sawtimber-sized logs) to bioenergy plants as feedstocks. However, they were well aware of the special challenges they face, and many expressed an insufficient understanding of bioenergy and its potential feedstocks to make any changes to their current management strategies specifically for bioenergy. There was a general feeling of risk aversion and resistance to shorter rotations, especially if they were already on their way to producing sawtimber.

Some Black farmers had tried other experimental crops, such as kenaf and stevia, without success, making them more skeptical of trying other crops for feedstocks, including energy grasses. Some said they might plant part of their acreage in energy grasses such as switchgrass or Miscanthus, or they might intercrop some pines with them, but none had enough faith in the continuation of the market to completely change their current farming strategies. However, one White landowner saw a lot of potential in energy grasses, particularly as a way to lift rural, majority Black communities and Black-landowning families out of poverty. He said:

> I hate to say this, but the quickest way to poverty is pine trees. There is a cash event only once every 16 years. The way to create wealth is mining or agriculture. To get other counties to be prosperous, they need to get out of pines. They need infrastructure and small businesses. Look at [three county names]. They all are based on timber, and they are all very poor. Timber is very cheap now. If 25% of the land in [county name] went to energy grasses, people would bring in money every month. There would also be dealerships, fertilizer sales, tractor sales. A grocery store. There would be more money in the county. People would have nicer houses and drive nicer cars. Pine trees are not doing it, and they continue to deteriorate . . . . Look at [county name]. It is 80% minority and very poor. They have a great county agent there. There could be a mill in [town name], which is on the railroad. Find absentee landowners and minority farmers and identify tracts. Start there with government initiatives to revitalize the agricultural base in that county. That would get the money flowing, and there would not need to be a subsidy, except for maybe at first. The landowners can prosper, and also other non-agricultural businesses. [County name] could have a crude oil refinery or distillery, and there could be other terminals surrounding [county name]. These should be located closer to the pipeline. Most of these refineries would be south of [highway].

He then drew us a map, after explaining his own imaginary for a future that is more economically prosperous, socially just, and contributing to a renewable future. It should be noted, however, that he had a financial stake in promoting expanded production of energy grasses.

ENVIRONMENTAL JUSTICE: FAIR TREATMENT AND MEANINGFUL INVOLVEMENT

The Environmental Protection Agency (EPA) defines environmental justice as "the fair treatment and meaningful involvement of all people regardless of race, color, national origin, or income with respect to the development, implementation and enforcement of environmental laws, regulations and policies" (EPA 2020). In the United States, communities of color often shoulder a disproportionate burden of the negative side effects of development, particularly when environmental regulations do not adequately safeguard human health (or when these regulations are blatantly ignored). While this burden is the result of a combination of factors, including local towns providing economic incentives for companies who place polluting industries in low-income areas, these are rooted in racist disregard for the health and safety of minorities.

This situation is not unique to the South; communities of color across the country are more likely to live in close proximity to sources of water and air pollution, including factories, highways, and congested inner cities, thus bearing a heavier burden of industrial development (Ahlerhs 2016; Shrader-Frechette 2002; Bullard 2005, 2011; Checker 2005; Outka 2012). Research on the links between bioenergy development and environmental justice has shown that similar dynamics are in play with biomass facilities in the U.S. South (Koester and Davis 2018; Dwivedi and Alavalapati 2009; Schelhas, Hitchner, and Brosius 2017, 2018; Hitchner, Schelhas, and Brosius 2017; Hitchner et al. 2014; Hitchner and Schelhas 2012) and other regions of the United States (Belyokav 2019; Collin and Collin 201; Kulcsar, Selfa, and Bain 2016). Large-scale biomass facilities can have disproportionate health effects on poor, non-White communities, resulting in disparities between those people who benefit from renewable energy and those who experience its side effects. Sometimes these health concerns are framed as NIMBY arguments, and opposition to biomass facilities does sometimes originate with White community members or with health or environmental organizations with a mostly White membership. Sometimes, White faces are used in public campaigns against biomass, as seen in figure 5.2. This could be done in order to make the problem seem more universal, or more relatable to people who would not otherwise be exposed to the negative environmental consequences of development.

As we discussed in chapter 1, a number of health and public safety organizations have made statements against the burning of biomass for energy, and much of the concern among bioenergy advocates about the "social acceptability" of bioenergy stems from these accusations that biomass is a threat to public health. As we noted in chapter 3, a number of people in Soperton, the majority of them Black, expressed concerns about the safety of the Range Fuels/LanzaTech plant, and about the pollution that community members (especially children attending the new school) would be exposed to through the burning of biomass or the increased traffic of logging trucks. Similar concerns were voiced in other sites as well. Community organizations that advocate for health and safety of minority populations, particularly when they are composed entirely or mostly of people of color, face especially vehement racism by some White community members, as evidenced in the quote containing racial slurs about one such group in one of our field sites.

Figure 5.2   Road sign promoting opposition to a biomass facility in Georgia. *Source*: Credit: Sarah Hitchner.

However, the main equity concerns regarding renewable energy development run deeper than those related to smoke, smog, or risks of explosion; these are concerns regarding structural racism. Mittlefehldt (2018, 876) puts this into broader historical context by noting that

> New energy technologies designed to run on renewable fuels became part of an older fossil fuel-based power structure that maintained deep historical inequalities. This power structure was reinforced by an environmental regulatory framework that focused more on future ambient air quality standards and less on the full participation of vulnerable populations in decision-making processes.

In this situation, the renewable energy company (based in Flint, Michigan, a place practically synonymous with failed environmental regulations) framed itself as an important actor in a developing field of technology that would reduce dependence on foreign oil and promote economic growth in an economically depressed city. Mittlefehldt (2018, 254) also considers race to be "one of the key cultural and political dimensions of technological systems," one that cannot be ignored in Flint or elsewhere.

This assertion by renewable energy companies and advocates that they can offer solutions to marginalized communities becomes problematic when seen as part of a larger pattern of systemic racism. This type of strategic positioning stretches across many realms of governance in which the powerful actors seek to be seen as beneficial while also maintaining their hold on power, thus controlling both the regulatory process and the narrative. This paternalistic, almost colonialist, mindset is perpetuated in numerous venues, from board-rooms to Congress, in which the racial and ethnic makeup of groups does not accurately reflect the wider community. Lack of representation, and more importantly, lack of true decision-making power, is an obstacle to free and fair participation in issues that directly affect communities of color. Concern about representation in decision-making processes on a number of scales is a key component of energy justice, which is not limited to equitable access to energy products.

Energy justice, as defined by Sovacool and Dworkin (2015, 436), is "a global energy system that fairly disseminates both the benefits and costs of energy services, and one that has representative and impartial decision-making." Other scholars have described energy justice as an emerging field which draws from environmental justice in its focus on the three tenets of distributional justice, procedural justice, and recognition justice (Sovacool et al. 2004; Sovacool et al. 2016; Walker et al. 2014; Holifield et al. 2018; Jenkins et al. 2016; Bickerstaff et al. 2013; Mittlefehldt 2016; Lavenda et al. 2021). Sovacool and Dworkin's definition of energy justice focuses on the first two, while recognition justice "requires that policies and programs must meet the standard of fairly considering and representing the cultures, values, and situations of all affected parties" (Whyte 2011, 201). Jenkins et al. (2021, 18) remind us of the inevitable costs of all forms of energy production and consumption and the multi-scalar impacts of these, emphasizing that

> Cleaner and lower carbon energy may be a human right, but securing it currently forces trade-offs with other human rights, leading to "green on green" and even "poor on poor" conflict. We must avoid conceptual approaches or research designs that obscure or mask this emerging spatial divide to energy justice.

Attention must also be paid, they note (ibid. 2021, 18) to the "intersectional injustices" arising from "existing inequalities of ethnicity, race, class, and social status." Energy justice, they note (ibid. 2021, 20), can reframe energy systems "not only as matters of national security, economic competitiveness, or environmental degradation, but as matters of social injustice" and that energy justice can "become a lever for *action* and community mobilization, so that new transformations to global energy systems can be intently debated, evaluated according to justice principles, and enacted."

As we have seen throughout this chapter, the racial disparities are numer-ous, and the racism remains deeply entrenched. It is a fact of daily life in the rural South, as it is elsewhere in the nation. When a new, experimental technology comes to town and makes big promises, people, especially those with a long history of disenfranchisement, are naturally skeptical. As we have seen, they take note of how companies try or do not try to engage with the community, fulfill promises of equal employment opportunities, and dispel fears about their facility (and actually do what they can to reduce its negative impacts on the community). People of color in these rural communities see how even when companies do come with good intentions, they are often thwarted by a local political system that is still run by elite Whites who feel threatened about "them taking over." For them, the fact that the companies and the local community leaders work together is often proof enough that things are unlikely to change for the better.

Bioenergy development thus has the potential to not only expose, but to exacerbate, existing racial inequalities. Mittlefehldt (2018, 876) notes:

> Existing political institutions, material infrastructure, and racial disparities shaped the development of renewable energy technologies. In places where environmental racism was deeply entrenched, industrial-scale renewable energy technologies maintained racial injustices. Without careful attention to how different types and scales of renewable technology applications interact with existing sociopolitical dynamics and racial divisions, future energy decisions are likely to reproduce the inequalities of the past.

Attention to the deeper structural issues of environmental justice, which are not just about unequal exposure to pollution, underscores the complicated relationships between race and bioenergy in the U.S. South. It requires focus on the historical and current racial dynamics in communities in which new bioenergy facilities are built, as well as an analysis of the configurations of power, influence, and social and natural capital that lead companies to choose particular locations. It also forces us to ask questions about how the experiences and perspectives of Black (and other non-White) Americans in heavily forested, economically depressed rural communities fit into, or complicate, the main narratives and counternarratives about bioenergy. Are their specific values, goals, concerns, and challenges accounted for within the larger pro-bioenergy imaginary, and what happens when these fall outside the bounds of it? Do these concerns reconfigure the overall imaginary, or do they contribute to something new?

We are not sure about the answers to these questions, but we do know that based on what we have witnessed across the South in communities where bioenergy facilities were located is that the concerns of minorities—about disparities in exposures to health and safety hazards, opportunities for quality education and well-paying jobs, access to wood markets, and the ability to overcome problems created by drug addiction, incarceration, and intimidation or brutality by police officers—were often sidelined, as these realities throw a wrench into the rosy portrait of bioenergy as a potential savior of rural America. But a related, and overarching, concern was that there was often a dearth of minorities in positions of decision-making power, making meaningful participation in bioenergy development processes illusory.

NOTES

1. These tribes later became known as the "Five Civilized Tribes," as they adopted European customs, including speaking English, converting to Christianity, and adopting private land ownership patterns.

2. The commodification of Black history, and its appropriation by Whites, was a theme that we encountered in several of our field sites. One White man remarked on the music festivals in the Mississippi Delta region: "There are blues festivals over there. Of course, they're 99% White now. There's blues tourism now, and sharecropper tourism. People come to see that."

3. LGBT civil rights advocacy group: https://www.hrc.org/

4. Gay and Lesbian Alliance Against Defamation: https://www.glaad.org/

5. The woman misunderstood what I (Hitchner) said about biofuels. I introduced our research on bioenergy and mentioned that some people opposed bioenergy in general because they felt its production competed with food crops, while others felt that wood-based bioenergy avoided this issue by using non-edible feedstocks.

6. One Black man felt strongly about drugs, and his potential solution, influenced by time spent overseas in the military in a nation with draconian laws, revealed his own racial biases:

It's good to put soldiers in the street, where everyone can see them. Everyone benefits. I'd like to work with local police to help small communities. Put soldiers on the streets in small communities instead of spending all that military money overseas. I bet you'd see a 50% drop in drugs, maybe 100%. This drug war . . . the county must run like a family . . . If my best friend were to give my kids drugs, he's not my friend anymore. You see? Americans stole America. It's hard to control it. The only way the U.S. is going to survive is if we bomb Mexico. Do you agree? All the drugs, they mess up people.

Conclusion

A New Bioenergy Imaginary in the U.S. South

Though the research for this book began almost ten years ago, our findings have continued relevance to an interrelated set of issues facing society, including energy transitions, climate change, forest conservation, and racial justice. Bioenergy has been promoted as a way to simultaneously meet multiple national objectives including rural development, energy independence, and climate change mitigation; the real impacts, however, are messy and complicated. This book situates the story of wood-based bioenergy in the U.S. South in the physical, social, political, and cultural landscapes of the region and links energy issues to those related to the history and potential futures of the region's forested landscape; the complex racial dynamics of the region, particularly as related to the biophysical landscape; the varying perceptions of and reactions to climate change; and the place of wood-based bioenergy in these discussions.

This book tells a story of the development and implementation of a new energy technology based on woody biomass in the U.S. South, with a particular focus on the encounter that took place as promotion of this technology at the national and regional levels met complex local contexts in the process of implementation. As Jasanoff and Kim (2009, 2013) have noted, such energy transitions are not simply a linear progression of technological development, but are complex social processes that are rooted in specific communities and situated within broader discourses, policies, and markets. We have drawn on three social science concepts as lenses for our analyses: imaginaries, scale, and translation.

The concept of socio-technical imaginaries helps us to understand the content and power of the bioenergy vision that emerged in the early 2000s, as well how it engendered specific types of bioenergy development in the U.S. South. Bioenergy has been promoted as a technology that, when

187

implemented, will produce a suite of economic, social, and environmental benefits that resonate with many people and interests. Seen as a socio-technical imaginary, it is a vision of the future in which the development and implementation of new energy technologies will result in energy independence, renewable and sustainable energy to address climate change, and rural development through the production of woody biomass. As such, it is a broad vision of a desirable future that can be embraced by various interest groups and has the power to shape policies and direct public and private expenditures toward its realization. Correspondingly, a favorable environment for bioenergy development was created as government policies were put in place to require certain levels of biofuel blending and bioenergy feedstocks, state and federal loan guarantees and other incentives were extended for construction of new plants, new commercial ventures were established to raise private funds for bioenergy development, and state and local development agencies matched these new companies with communities hungry for industrial development. The power of this imaginary and the government and industry support behind it were reflected in the near-frenzy of bioenergy development that ensued, which in some cases got well ahead of fully vetted technologies.

While the vision of new technologies providing multiple social, economic, and environmental benefits was enough to create the enthusiasm and conditions for bioenergy development, that was just the beginning of the story. As we noted in chapter 2, recent work on imaginaries calls attention to the ways that they are repeated, contested, and altered as they are taken up by "ordinary people" and the need for researchers to locate imaginaries in concrete actors, social contexts, and material conditions (Smith and Tidwell 2016; Strauss 2006). Through our ethnographic research, we have endeavored to do that. We began by highlighting the metaphors, cultural models, and conventional discourses that we heard at the local level in response to bioenergy development. Our analysis was enhanced by the additional lenses of scale and translation, as we examined how the national bioenergy imaginary was contested and reshaped as it encountered local contexts over time. Concerns, interests, and visions for bioenergy differed both across and within scales, and there were significant power differentials throughout associated with both financial and discursive resources. The bioenergy imaginary originated at the top, with little opportunity for input from less powerful people. Yet imaginaries are not accepted wholesale and may speak to people in different ways, and local people drew on a wide range of metaphors and conventional discourses to reshape the bioenergy imaginary to suit their lives and interests. In the end, the bioenergy imaginary was translated into multiple imaginaries, broader than bioenergy, that sometimes overlapped and sometimes were incommensurable.

Forested landscapes and forest-dependent communities in the U.S. South are the products of a long history of changing forest product use and forest management. While outside interests variously see forests as sources of raw materials, stored carbon, and repositories of wildlife and biodiversity, they are also home for many people who earn income from forests, appreciate the aesthetics of a forested landscape, and patriotically embrace many national concerns and issues. The national bioenergy imaginary and associated development fit awkwardly with these local and regional interests and imaginaries of forests. One disconnect occurred when researchers and foresters responded to the bioenergy imaginary with new kinds of trees and silvicultural systems, while landowners embraced stories of new markets for currently "wasted" wood to add value to their current pine forest systems. Another disconnect occurred when changing forest management options were seen as impinging on other local forest values, such as aesthetics, wildlife, and hunting. And while bioenergy promised new markets, landowners responded slowly due to the long-term nature of forestry and experiences with other promised markets that had failed to materialize.

People in the rural South have participated in and been influenced by recent trends in distrust of government and hostility toward science. The politicization of climate change, with disbelief widespread in the rural U.S. South, made bioenergy's potential to address climate change more of a liability than an attraction for many people. The general distrust of government and its initiatives combined with disbelief in climate change to feed local discourses about bioenergy being yet another government initiative devised for the enrichment of a few businesses. This jaded view was in turn fed by uncertainties in technologies and the failure of bioenergy development to produce promised benefits such as jobs and new markets for wood products. Yet, development interests in local communities saw wood-based industries as the best hope for their communities; they recognized that industrial development is often subsidized and incentivized, and they retained hope and faith that bioenergy would become viable. The effects of long-standing racial relationships and inequities continue to manifest themselves in communities we worked in, raising questions about who will have opportunities to participate in forestry and how the benefits and costs of bioenergy will be distributed—development for what and for whom.

The ethnographic analysis of bioenergy development in the U.S. South through the pillars of energy, forests, climate, and race allows us to sort through the messiness of local contexts to elucidate the underlying dynamics that lie at the intersection of the lenses of imaginaries, scale, and translation. In practical terms, this analysis can provide guidance to promote renewable energy policy and production in ways that are beneficial to local communities most directly affected by these developments. Equally important, this sort of ethnographic approach might further our efforts to identify areas of convergence and points of compromise in a charged political atmosphere characterized by division, confrontation, and polarization.

As we noted in our introductory chapter, translating the reality of bioenergy development across multiple imaginaries and spheres of influence is a distinct challenge and forces difficult conversations about incommensurable ends when some groups stand to directly benefit from bioenergy development at the expense of others. Attention to the dynamics at play in the ecological, social, cultural, and political spheres is vital when assessing possible impacts of future bioenergy development trajectories on local communities. As a region, the U.S. South has a unique history and cultural landscape, and the stories embedded in these forests and watersheds have much to contribute to national conversations about numerous pressing contemporary issues.

Many of the issues we have discussed in this book have been at the forefront of the news, leading to societal and policy advances as well as backlashes. Debates continue about the appropriate role of government, the nature of our country and democracy, the need to address past racial discrimination and appropriate mechanisms to do so, and the existence and importance of addressing climate change. These debates influence bioenergy development and reflect many of the processes we have discussed in this book. They put forward various imagined futures and are communicated through metaphors and conventional discourses that are in part national but are reshaped and reimagined in complex ways across scales and interest groups. They are also reflected in policy changes and investments, which generate both hope and despair as they are implemented and contested. The search for common ground—and new imaginaries—that are sufficient for progress in creating a safe, productive, and just world for future generations will not be easy. Energy, climate change, racial equity, and forested landscapes are likely all to have enduring central roles. Understanding the messy multi-scalar processes that lie ahead is critical, and will require robust conceptual frameworks and careful empirical analysis.

As we complete this book, the emerging U.S. leadership is focusing on the connections between—and vital importance of—climate, forests, energy, and race and the role of science in addressing them. On January 27, 2021, President Joe Biden signed an executive order (White House 2021) the purpose of which, putting climate front and center on the national agenda, is as stated:

> This order builds on and reaffirms actions my Administration has already taken to place the climate crisis at the forefront of this Nation's foreign policy and national security planning, including submitting the United States instrument of acceptance to rejoin the Paris Agreement. In implementing—and building

upon—the Paris Agreement's three overarching objectives (a safe global temperature, increased climate resilience, and financial flows aligned with a pathway toward low greenhouse gas emissions and climate-resilient development), the United States will exercise its leadership to promote a significant increase in global climate ambition to meet the climate challenge.

The executive order also addresses environmental and climate justice, as it strengthens the government's commitments to address racial, ethnic, and socioeconomic disparities in exposure to industrial toxins by ordering the creation of "a geospatial Climate and Economic Justice Screening Tool and shall annually publish interactive maps highlighting disadvantaged communities." It also vows to "strengthen enforcement of environmental violations with disproportionate impact on underserved communities" and "create a community notification program to monitor and provide real-time data to the public on current environmental pollution, including emissions, criteria pollutants, and toxins, in places with the most significant exposure to such pollution."

This executive order places renewable energy technologies at the forefront of new energy initiatives, aiming for "a carbon pollution-free electricity sector no later than 2035," as well as promising to "pause new oil and natural gas leases on public lands or in offshore waters," "to eliminate fossil fuel subsidies from the budget request for Fiscal Year 2022 and thereafter," and to ensure that "federal funding is used to spur innovation, commercialization, and deployment of clean energy technologies and infrastructure." It recognizes that "America's farmers, ranchers, and forest landowners have an important role to play in combating the climate crisis and reducing greenhouse gas emissions, by sequestering carbon in soils, grasses, trees, and other vegetation and sourcing sustainable bioproducts and fuels." This order, then, clearly reflects the interrelatedness of the issues of climate change, forests, race, and energy.

This order also echoes the pro-bioenergy imaginary that simultaneously addresses renewable energy, sustainability, energy security, climate change, rural development, and environmental justice we have been discussing throughout this book by stating:

> To secure an equitable economic future, the United States must ensure that environmental and economic justice are key considerations in how we govern. That means investing and building a clean energy economy that creates well-paying union jobs, turning disadvantaged communities—historically marginalized and overburdened—into healthy, thriving communities, and undertaking robust actions to mitigate climate change while preparing for the impacts of climate change across rural, urban, and Tribal areas. Agencies shall make achieving environmental justice part of their missions by developing programs, policies,

and activities to address the disproportionately high and adverse human health, environmental, climate-related and other cumulative impacts on disadvantaged communities, as well as the accompanying economic challenges of such impacts. It is therefore the policy of my Administration to secure environmental justice and spur economic opportunity for disadvantaged communities that have been historically marginalized and overburdened by pollution and underinvestment in housing, transportation, water and wastewater infrastructure, and health care.

This can be seen as a new socio-technical imaginary, but one that is more explicitly social and inclusive than earlier bioenergy imaginaries. It remains to be seen how it will unfold and develop in the coming years as it enters the world of concrete actors, social contexts, material conditions, new technologies, and fluctuating energy markets. This executive order, and the legislation that will come from it, helped to steer the discussions at United Nations environmental meetings. In November 2021, Glasgow, Scotland, hosted the UNFCCC 26th Conference of the Parties (COP 26), rescheduled from 2020, when the venue was converted to a field hospital after a surge in COVID-19 cases in Scotland (Practical Law Environment 2020). Climate targets were in the spotlight, with other issues such as environmental justice, clean energy development, and maintenance and increase of global forest cover treated as integral, not just peripheral, concerns.

What happens at these highest levels of governance (at national and international levels) will likely continue to translate to real developments in local communities in the U.S. South. The Georgia Biomass pellet mill in Waycross is still operating at full capacity, shipping pellets to Europe. The Freedom Pines Biorefinery in Soperton, Georgia, is still in operation as well. As of early 2021, a subgroup of LanzaTech formed, called LanzaJet, who will focus production on aviation fuels made primarily from southern yellow pine feedstock around Soperton. LanzaJet now has commitments from British Airways, who will purchase SAF (sustainable aviation fuels) from the facility, in an effort to meet its nation's own climate targets. British Airways "expects the fuel to be available to power a number of its flights by the end of 2022," working toward a goal to "decarbonise and achieve net zero by 2050" (Burton 2021).

These communities, and many others in the U.S South with a long history of integration with an agricultural and forested landscape, as well as deep racial divides and socioeconomic vulnerabilities, will continue to be sought as sites for wood-based bioenergy. These areas, as we have explored in this book, are steeped in a history of economic dependence on forest products, and on exploitative systems of labor, with people accustomed to having control struggling to maintain it. As we have seen, minority members of these communities are, in some cases, pushing against the confines of entrenched power imbalances and carving out new opportunities. Many White people (certainly not all) are also grappling with the nation's history of racial injustice through awareness, inclusivity, and development of concrete programs to address the consequences of this long history of discrimination.

This book has explored the myriad points of friction that complicate the pro-bioenergy imaginary in which wood-based bioenergy simultaneously meets a number of goals by promoting energy independence, enhancing forest health, strengthening local wood markets, mitigating climate change, and benefitting rural communities, in particular the most oppressed and vulnerable members of those communities. Not all actors agree that wood-based bioenergy will help to address climate change. Quite a number of people in rural communities are not convinced that anthropogenic climate change is something to worry about or even real at all, although record temperatures, new rainfall patterns, and frequent forest fires may influence beliefs. Nor do all actors agree that forests can be used for fuel in a sustainable manner, especially when local resources are shipped overseas to provide renewable energy for someone else. It cannot be assumed that all forest landowners and managers will be willing to grow and harvest trees in the manner that the forest economists agree is necessary to supply biomass on a large enough scale to feed these developing facilities. Likewise, not all actors in local communities are eager to embrace new experimental technologies in their hometowns, given the environmental, health, and safety risks that they perceive come along with it. Companies are sometimes unprepared for the economic and social realities of the communities in which they locate their plants. They may see plenty of trees to provide feedstock and plenty of un- or underemployed people to provide labor (and they may be offered all kinds of incentives by states, counties, and cities to locate there), but the ability to fully utilize those feedstocks or employ those people may be limited. For local community members, who have long been exposed to racial and economic oppression, neglect by the state, and cycles of enthusiasm and disillusionment with other industries (including forest-based industries), the process of bioenergy development is an extension of unbalanced power dynamics that can lead to further disenfranchisement.

Here we find that our role as anthropologists is not simply that of a conduit of information—although we do write down what people say, as closely as possible to just how they said it; rather, we are ourselves translators. It is our task to draw attention to the ways that imaginaries get complicated on the ground and to what happens when elements of these imaginaries and realities are simply, in the end, incommensurable. Having acknowledged that, we believe that there is an intrinsic value in acknowledging incommensurability rather than seeking closure in elegant models built on abstractions. In the book *Friction*, Anna Tsing (2005, 274) famously observed that "collaboration was not consensus making but rather an opening for productive confusion." Similarly, accepting the principle of incommensurability is not a dead end and does not foreclose the possibility of communication, collaboration, or reconciliation. Indeed, it invites these. In our view, these zones of incommensurability are a powerfully generative space for reimagining how we might respond to the complex sustainability challenges facing our country and the planet.

Bibliography

25x'25. 2010. 25x'25 "Meeting the Goal: A Progress Report." www.25x25.org.

Abrams, Mark D., and Gregory J. Nowacki. 2008. "Native Americans as Active and Passive Promoters of Mast and Fruit Trees in the Eastern USA." *Holocene* 18, no. 7: 1123–37. doi: 10.1177/0959683608095581.

Abt, Robert C., Karen Lee Abt, Fred W. Cubbage, and Jesse D. Henderson. 2010. "Effect of Policy-based Bioenergy Demand on Southern Timber Markets: A Case Study of North Carolina." *Biomass and Bioenergy* 34, no. 12: 1679–86. doi: 10.1016/j.biombioe.2010.05.007.

Aguilar, Francisco X. 2012. "Green Economy." UNECE (United Nations Economic Commission for Europe)/FAO (Food and Agricultural Organization of the United Nations. Forestry and Timber Section, Palais des Nations, CH-1211 Geneva 10, Switzerland, http://www.unece.org/fileadmin/DAM/timber/wood_energy/wood -energy-policy-brief-2012.pdf

Aguilar, Francisco X. 2014. "Wood Energy in the E.U. and U.S.: Assessment and Outlook to 2030." In *Wood Energy in Developed Economies: Resource Management, Economics, and Policy*, edited by Francisco X. Aguilar, 306–27. London: Earthscan.

Aguilar, Francisco, and H. E. "Gene" Garrett. 2009. "Perspectives of Woody Biomass for Energy: Survey of State Foresters, State Energy Biomass Contacts, and National Council of Forestry Association Executives." *Journal of Forestry* 107, no. 6: 297–306. doi: 10.1093/jof/107.6.297.

Aguilar, Francisco X., Marissa "Jo" Daniel, and Lara Narine. 2013. "Opportunities and Challenges to the Supply of Woody Biomass for Energy from Missouri Nonindustrial Privately Owned Forestlands." *Journal of Forestry* 111, no. 4: 249–60. doi: 10.5849/jof.13-009.

Aguilar, Francisco X., and Warren Mabee. 2014. "Wood: A Renewable Source of Energy." In *Wood Energy in Developed Economies: Resource Management, Economics, and Policy*, edited by Francisco X. Aguilar, 1–31. London: Earthscan.

Ahlers, Christopher D. 2016. "Race, Ethnicity, and Air Pollution: New Directions in Environmental Justice." *Environmental Law* 46, no. 4: 713–58.

Aiken, Charles S. 1998. *The Cotton Plantation South since the Civil War.* Baltimore, MD: Johns Hopkins University Press.

Alavalapati, Janaki R. R., Pankaj Lal, Andres Susaeta, Robert C. Abt, and David N. Wear. 2013. "Forest Biomass-Based Energy." In *The Southern Forest Futures Project: Technical Report,* edited by David N. Wear and John G. Greis, 213–55. Gen. Tech. Rep. SRS-178. Asheville, NC: U.S. Department of Agriculture Forest Service, Southern Research Station.

Álvarez, Román, and M. Carmen Africa Vidal, eds. 1996. *Translation, Power, Subversion.* Clevedon: Multilingual Matters Ltd.

Anderson, Benedict. 1983. *Imagined Communities: Reflections on the Origin and Spread of Nationalism.* London: Verso.

Anderson, Eugene N. 2014. *Caring for Place: Ecology, Ideology, and Emotion in Traditional Landscape Management.* New York: Routledge.

Anonymous. 2008. "Stop Giving Money to People Who Hate Us: Anti-Oil Mud Slinging Festival." *DIY Electric Car Forum.* July 2008. http://www.diyelectriccar .com/forums/showthread.php/stop-giving-money-people-hate-us-19762.html.

Ansolabehere, Stephen, and David M. Konisky. 2014. *Cheap and Clean: How Americans think about Energy in the Age of Global Warming.* Cambridge, MA: MIT Press.

Argus. 2014. "Georgia Biomass 750 000 T/Y Pellet Plant for Sale." 18 June 2014. http://www.conbio.info/post/argus-georgia-biomass-750-000-ty-pellet-plant-for -sale/.

Aryee, Alberta N. A., Benjamin K. Simpson, Roger I. Cue, and Leroy E. Phillip. 2011. "Enzymatic Transesterification of Fats and Oils from Animal Discards to Fatty Acid Ethyl Esters for Potential Fuel Use." *Biomass and Bioenergy* 35, no. 10: 4149–57. doi: 10.1016/j.biombioe.2011.06.002.

Asad, Talal. 1986. "The Concept of Cultural Translation in British Social Anthropology." In *Writing Culture: The Poetics and Politics of Ethnography,* edited by James Clifford and George E. Marcus, 141–64. Berkeley, CA: University of California Press.

Bailey, Conner, Becky Barlow, and Janice Dyer. 2019. "Practical Constraints to Timber Management among African American Owners of Heirs' Property." *Landscape and Urban Planning* 188: 180–87. doi: 10.1016/j.landurbplan.2019.03.008.

Bailey, Conner, Janice F. Dyer, and Larry Teeter. 2011. "Assessing the Rural Development Potential of Lignocellulosic Biofuels in Alabama." *Biomass and Bioenergy* 35: 1408–17. doi: 10.1016/j.biombioe.2010.11.033.

Bailey, Conner, Peter Sinclair, John Bliss, and Kami Perez. 1996. "Segmented Labor Markets in Alabama's Pulp and Paper Industry." *Rural Sociology* 61, no. 3: 475–96. doi: 10.1111/j.1549-0831.1996.tb00630.x

Bailey, Ian, and Hoayda Darkal. 2018. "(Not) Talking About Justice: Justice Self-Recognition and the Integration of Energy and Environmental-Social Justice into Renewable Energy Siting." *Local Environment* 23, no. 3: 335–51. doi: 10.1080/13549839.2017.1418848.

Bailis, Robert, and Jennifer Baka. 2011. "Constructing Sustainable Biofuels: Governance of the Emerging Biofuel Economy." *Annals of the Association of American Geographers* 101, no. 4: 827–38. doi: 10.1080/00045608.2011.568867.

Bain, Carmen, and Theresa Selfa. 2013. "Framing and Reframing the Environmental Risks and Economic Benefits of Ethanol Production in Iowa." *Agriculture and Human Values* 30, 351–64. doi: 10.1007/s10460-012-9401-y.

Baker, Allen, and Steven Zahniser. 2006. "Ethanol Reshapes Corn Market." USDA-Economic Research Service. http://www.ers.usda.gov/AmberWaves/April06/Features/Ethanol.htm.

Ballo, Ingrid Foss. 2015. "Imagining Energy Futures: Sociotechnical Imaginaries of the Future Smart Grid in Norway." *Energy Research & Social Science* 9: 9–20. doi: 10.1016/j. erss.2015.08.015

Barney, Jacob N., and Joseph M. di Tomaso. 2011. "Global Climate Niche Estimates for Bioenergy Crops and Invasive Species of Agronomic Origins: Potential Problems and Opportunities." *Plos One* 6, no. 3: e17222. doi: 10.1371/journal .pone.0017222.

Becker, Dennis R., Casandra Moseley, and Christine Lee. 2011. "A Supply Chain Analysis Framework for Assessing State-Level Forest Biomass Utilization Policies in the United States." *Biomass and Bioenergy* 35, no. 4: 1429–39. doi: 10.1016/j. biombioe.2010.07.030.

Behringer, Wolfgang. 2010. A *Cultural History of Climate*. Cambridge, UK: Polity Press.

Belyakov, Nikolay. 2019. *Sustainable Power Generation: Current Status, Future Challenges, and Perspectives*. San Diego, CA: Academic, an Imprint of Elsevier.

Bengston, David N., Stanley T. Asah, and Brett J. Butler. 2011. "The Diverse Values and Motivations of Family Forest Owners in the United States: An Analysis of an Open-Ended Question in the National Woodland Owner Survey." *Small-scale Forestry* 10: 339–55. doi: 10.1007/s11842-010-9152-9.

Berlik, Mary M., David B. Kittredge, and David R. Foster. 2002. "The Illusion of Preservation: A Global Environmental Argument for the Local Production of Natural Resources." *Journal of Biogeography* 29: 1557–68. doi: 10.1046/j.1365-2699.2002.00768.x.

Bettis, Jerry L, Sr., and Jerry L. Bettis, Jr. 2017. "Perceptions of African Americans' Persistence in Careers in the Forestry/Natural Resources Management Professions: The Tuskegee University Case." *NACTA Journal* 61, no. 1: 77–82.

Beyea, Jan, Wayne A. Hoffman, and James H. Cook. 1994. "Species Diversity in Large-Scale Energy Crops and Associated Policy Issues." Annual Report for Subcontract No. IBX-SL237C with Martin Marietta Energy Systems. National Audubon Society, New York.

Bickerstaff, Karen, Gordon Walker, and Harriet Bulkeley. 2013. *Energy Justice in a Changing Climate: Social Equity and Low-Carbon Energy*. London: Zed Books Ltd.

Bidwell, David. 2013. "The Roles of Values in Public Beliefs and Attitudes Towards Commercial Wind Energy." *Energy Policy* 58: 189–99. doi: 10.1016/j. enpol.2013.03.010.

Bliss, John C., Tamara L Walkingstick, and Conner Bailey. 1998. "Development of Dependency? Sustaining Alabama's Forest Communities." *Journal of Forestry* 96, no. 3: 24–30. doi: 10.1093/jof/96.3.24.

Bliss, John C., and Conner Bailey. 2005. "Pulp, Paper, and Poverty: Forest-based Rural Development in Alabama, 1950–2000." In *Communities and Forests: Where People Meet the Land*, edited by R.G. Lee and D.R. Field, 138–58. Corvallis, OR: Oregon State University Press.

Bliss, John C., and Warren Flick. 1994. "With a Saw and a Truck: Alabama's Pulpwood Producers." *Forest and Conservation History* 38: 79–89. doi: 10.2307/3983722.

Bomgardner, Melody M. 2012. "LanzaTech Buys Range Fuels Facility: Revival Plan: Waste Woody Biomass Would Be Transformed into Jet Fuel." *Chemical and Engineering News*, July 12, 2012. https://cen.acs.org/articles/90/web/2012/01/LanzaTech-Buys-Range-Fuels-Facility.html.

Botha, Tyron, and Harro von Blottnitz. 2006. "A Comparison of the Environmental Benefits of Bagasse-derived Electricity and Fuel Ethanol on a Life-cycle Basis." *Energy Policy* 34: 2654–61. doi: 10.1016/j.enpol.2004.12.017.

Bourgeon, Jean-Marc, and David Tréguer. 2010. "Killing Two Birds with One Stone: U.S. and E.U. Biofuel Programmes." *European Review of Agricultural Economics* 37, no. 3: 369–94. doi: 10.1093/erae/jbq025.

Box, Jason E., William T. Colgan, Torben Røjle Christensen, Niels Martin Schmidt, Magnus Lund, Frans-Jan W. Parmentier, Ross Brown, et al. 2019. "Key Indicators of Arctic Climate Change: 1971–2017." *Environmental Research Letters* 14, no. 4: 045010. doi: 10.1088/1748-9326/aafc1b.

Bracmort, Kelsi. 2015. "The Renewable Fuel Standard (RFS): Cellulosic Biofuels." Report 7-5700. Congressional Research Service, Washington, DC.

Brandeis, Consuelo, and Zhimei Guo. 2016. "Decline in the Pulp and Paper Industry: Effects on Backward-Linked Forest Industries and Local Economies." *Forest Products Journal* 66, no. 1–2: 113–18. doi: 10.13073/FPJ-D-14-00106.

Brenner, Neil. 2001. "The Limits to Scale? Methodological Reflections on Scalar Structuration." *Progress in Human Geography* 25, no. 4: 591–614. doi: 10.1191/030913201682688959.

Brisman, Avi, and Nigel Reece Walters. 2018. "Southernizing Green Criminology: Human Dislocation, Environmental Injustice and Climate Apartheid." *Justice, Power and Resistance* 2, no. 1: 1–21.

Broad, Kenneth, Anthony Leiserowitz, Jessica Weinkle, and Marissa Steketee. 2007. "Misinterpretations of the 'Cone of Uncertainty' in Florida during the 2004 Hurricane Season." *Bulletin of the American Meteorological Society* 88, no. 5: 651–68. doi: 10.1175/BAMS-88-5-651.

Brook, Barry W., Tom Blees, Tom M. L. Wigley, and Sanghyun Hong. 2013. "Silver Buckshot or Bullet: Is a Future "Energy Mix" Necessary?" *Sustainability* 10: 302. doi: 10.3390/su10020302.

Brosius, J. Peter. 2004. "What Counts as Local Knowledge in Global Environmental Assessments and Conventions?" Address to Plenary Session on "Integrating Local and Indigenous Perspectives into Assessments and Conventions," at conference "Bridging Scales and Epistemologies: Linking Local Knowledge and Global

Science in Multi-Scale Assessments." Biblioteca Alexandrina, Alexandria, Egypt, March 17–20, 2004. http://www.millenniumassessment.org/documents/bridging/papers/Brosius.peter.pdf.

Brosius, J. Peter, John Schelhas, and Sarah Hitchner. 2013. "Social Acceptability of Bioenergy in the U.S. South." In: *Proceedings of the Seventh National New Crops Symposium,* Washington, DC, 12–16 October.

Brosius, J. Peter, and Lisa M. Campbell. 2010. "Introduction: Collaborative Event Ethnography: Conservation and Development Trade-Offs at the Fourth World Conservation Congress." *Conservation and Society* 8, no. 4: 245–55. doi: 10.4103/0972-4923.78141.

Brown, Robert C., and Tristan R. Brown. 2012. *Why are We Producing Biofuels? Shifting to the Ultimate Source of Energy.* Ames, IA: Brownia.

Brown, Trent. 2016. "Sustainability as Empty Signifier: Its Rise, Fall, and Radical Potential." *Antipode,* 48: 115–33. doi: 10.1111/anti.12164.

Bryce, Robert. 2008. *A Gusher of Lies: The Dangerous Delusions of Energy Independence.* New York: Public Affairs.

Buford, Marilyn A., and Daniel G. Neary. 2010. "Sustainable Biofuels from Forests: Meeting the Challenge." Ecological Society of America Biofuels and Sustainability Reports. https://www.fs.usda.gov/treesearch/pubs/34785.

Bullard, Robert D., ed. 2005. *The Quest for Environmental Justice: Human Rights and the Politics of Pollution.* First edition. San Francisco, CA: Sierra Club Books.

Bullard, Robert D. 2011. "Dismantling Energy Apartheid in the United States." *Dissident Voice* http://dissidentvoice.org/2011/02/dismantling-energy-apartheid-in-the-united-states/.

Burnham, Morey, Weston Eaton, Theresa Selfa, Claire Hinrichs, Andrea Feldpausch-Parker. 2017. "The Politics of Imaginaries and Bioenergy Sub-niches in the Emerging Northeast U.S. Bioenergy Economy." *Geoforum* 82: 66–76. doi: 10.1016/j.geoforum.2017.03.022.

Burton, Freya. 2021. "British Airways Fuels Its Future with Second Sustainable Aviation Fuel Partnership." *Media News Release,* February 9, 2021. https://www.lanzatech.com/2021/02/09/british-airways-fuels-its-futures-with-second-sustainable-aviation-fuel-partnership/.

Büscher, Bram E., and Tendayi Mutimukuru. 2007. "Buzzing Too Far? The Ideological Echo of Global Governance Concepts on the Local Level: The Case of the Mafungautsi Forest in Zimbabwe." *Development Southern Africa* 24: no. 5: 649–64. doi: 10.1080/03768350701650512.

Butler, Brett J., and Earl C. Leatherberry. 2004. "America's Family Forest Owners." *Journal of Forestry* 102, no. 7: 4–9. doi: 10.1093/jof/102.7.4.

Butler, Brett J., Jaketon H. Hewes, Brenton J. Dickinson, Kyle Andrejczyk, Sarah M. Butler, and Marla Markowski-Lindsay. 2016. "Family Forest Ownerships of the United States, 2013: Findings from the USDA Forest Service's National Woodland Owner Survey." *Journal of Forestry* 14, no. 6: 638–47. doi: 10.5849/jof.15-099.

Caldas, Marcellus M., Jason S. Bergtold, Jeffrey M. Peterson, Russell W. Graves, Dietrich Earnhart, Sheng Gong, Brian Lauer, and J. Christopher Brown. 2014. "Factors Affecting Farmers' Willingness to Grow Alternative Biofuel

Feedstocks across Kansas." *Biomass and Bioenergy* 66: 223–31. doi: 10.1016/j. biombioe.2014.04.009.

Calver, Philippa, and Neil Simcock. 2021. "Demand Response and Energy Justice: A Critical Overview of Ethical Risks and Opportunities within Digital, Decentralised, and Decarbonised Futures." *Energy Policy* 151: 112198. doi: 10.1016/j.enpol.2021.112198.

Cama, Timothy. 2020. "The Anniversary of the Snowball Both Sides Love to Troll." *E&E News.* March 13. 2020. https://www.eenews.net/stories/1062588137.

Campbell, Lisa, Catherine Corson, Noella J. Gray, Kenneth Iain MacDonald, and J. Peter Brosius. 2014. "Studying Global Environmental Meetings to Understand Global Environmental Governance: Collaborative Event Ethnography at the 10th Conference of Parties to the Convention on Biological Diversity." *Global Environmental Politics* 14, no. 3: 1–20. doi: 10.1162/GLEP_e_00236.

Candea, Matei. 2007. "Arbitrary Locations: In Defense of the Bounded Field-site." *Journal of the Royal Anthropological Institute* 13, no. 1: 167–84.

Cardwell, Diane. 2012. "Military Spending on Biofuels Draws Fire." *New York Times.* August 27, 2012. https://www.nytimes.com/2012/08/28/business/military -spending-on-biofuels-draws-fire.html.

Carlson, Kenneth D., John C. Gardner, Vernon L. Anderson, and James J. Hanzel. 1996. "Crambe: New Crop Success." In *Progress in New Crops: New Opportunities, New Technologies*, edited by Jules Janick, 306–22. Alexandria, VA: ASHS Press.

Carroll, Wayne D., Peter R. Kapeluck, Richard A. Harper, and David H. Van Lear. 2002. "Historical Overview of the Southern Forest Landscape and Associated Resources." In *Southern Forest Resource Assessment*, edited by David N. Wear and John G. Greis, 583–605. General Technical Report SRS-53. Asheville, NC: U.S. Department of Agriculture Forest Service Southern Research Station.

Carter, Mason C., Robert C. Kellison, and R. Scott Wallinger. 2015. *Forestry in the U.S. South: A History.* Baton Rouge: Louisiana State University Press.

Castoriadis, Cornelius. 1987. *The Imaginary Institution of Society.* Cambridge, MA: MIT Press.

Chapman, Dan. 2012. "Warnings Ignored in Range Fuels Debacle." *Atlanta Journal-Constitution.* September 2, 2012. https://www.ajc.com/business/warnings-ignored -range-fuels-debacle/OGq6tJq3lEFyuXnZtFHZaO/.

Checker, Melissa. 2005. *Polluted Promises: Environmental Racism and the Search for Justice in a Southern Town.* New York: New York University Press.

Christian, Donald P., Wayne Hoffman, Joann M. Hanowski, Gerald J. Niemi, and G. Jan Beyea. 1998. "Bird and Mammal Diversity on Woody Biomass Plantations in North America." *Biomass and Bioenergy* 14: 395–402. doi: 10.1016/ S0961-9534(97)10076-9.

Chunxiang Fu., Jonathon R. Mielenz, Xiou Xirong, Ge Yaxin, Choo Y. Hamilton, Miguel Rodriguez Jr., Chen Fang Marcus Foston, Arthur Ragauskas, Joseph Bouton, Richard Dixon, and Zeng-Wu Wang. 2011. "Genetic Manipulation of Lignin Reduces Recalcitrance and Improves Ethanol Production from Switchgrass." *Proceedings of the National Academy of Sciences USA* 108, no. 9: 3803–08. doi: 10.1073/pnas.1100310108.

Collin, Robin Morris, and Collin, Robert W., editors. 2014. *Energy Choices: How to Power the Future*. Santa Barbara, CA: Praeger.

Cook, James, and Jan Beyea. 2000. "Bioenergy in the United States: Progress and Possibilities." *Biomass and Bioenergy* 18: 441–55. doi: 10.1016/S0961-9534(00)00011-8.

Cooksey, Elizabeth B. 2018. "Treutlen County." In *New Georgia Encyclopedia*. https://www.georgiaencyclopedia.org/articles/counties-cities-neighborhoods/treutlen-county.

Cornwall, Andrea, and Karen Brock. 2005. "Beyond Buzzwords: 'Poverty Reduction,' 'Participation,' and 'Empowerment' in Development Policy." Program Paper Number 10. United Nations Research Institute for Social Development (UNRISD), Geneva, Switzerland. http://www.unrisd.org/unrisd/website/document.nsf/(httpPublications)/F25D3D6D27E2A1ACC12570CB002FFA9A?OpenDocument.

Crate, Susan A. 2011. "Climate and Culture: Anthropology in the Era of Contemporary Climate Change." *Annual Review of Anthropology* 40:175–94. doi: 10.1146/annurev.anthro.012809.104925.

Crate Susan A., and Mark Nuttall, eds. 2009. *Anthropology and Climate Change: From Encounters to Actions*. Walnut Creek: Left Coast Press.

Creutzig, Felix, N.H. Ravindranath, Göran Berndes, Simon Bolwig, Ryan Bright, Francesco Cherubini, Helena Chum, et al. 2015. "Bioenergy and Climate Change Mitigation: An Assessment." *GCB-Bioenergy* 7: 916–44. doi: 10.1111/gcbb.12205.

Curran, Giorel. 2012. "Contested Energy Futures: Shaping Renewable Energy Narratives in Australia." *Global Environmental Change* 22: 236–44. doi: 10.1016/j.gloenvcha.2011.11.009.

Damery David, Jeffery Benjamin, M. Kelty, and R. J. Lilieholm. 2009. "Developing a Sustainable Forest Biomass Industry: Case of the US Northeast." http://www.forestbonds.net/sites/default/files/userfiles/1file/developing_sustainable_biomass_strategy_northeast_usa.pdf.

Daniel, Pete. 2013. *Dispossession: Discrimination against African American Farmers in the Age of Civil Rights*. Chapel Hill, NC: University of North Carolina Press.

de Gorter, Harry, and David R. Just. 2009. "The Welfare Economics of a Biofuel Tax Credit and the Interaction Effects with Price Contingent Farm Subsidies." *American Journal of Agricultural Economics* 91, no. 2: 477–88. doi: 10.1111/j.1467-8276.2008.01190.x.

Delaney, David, and Helga Leitner. 1997. "The Political Construction of Scale." *Political Geography* 16, no. 2: 93–97. doi: 10.1016/S0962-6298(96)00045-5.

Delcourt, Paul A., and Hazel R. Delcourt. 2004. *Prehistoric Native Americans and Ecological Change: Human Ecosystems in Eastern North America since the Pleistocene*. Cambridge: Cambridge University Press.

Delina, Laurence L. 2018. "Whose and What Futures? Navigating the Contested Coproduction of Thailand's Energy Sociotechnical Imaginaries." *Energy Research and Social Science* 35: 48–56. doi: 10.1016/j.erss.2017.10.045.

Delina, Laurence L., and Anthony Janetos. 2018. "Cosmopolitan, Dynamic, and Contested Energy Futures: Navigating the Pluralities and Polarities in the Energies of Tomorrow." *Energy Research and Social Science* 35: 1–10. doi: 10.1016/j. erss.2017.11.031.

Delshad, Ashlie B., Raymond, Leigh, Vanessa Sawicki, and Duane Wegener. 2010. "Public Attitudes Toward Political and Technological Options for Biofuels." *Energy Policy* 38: 3414–25. doi: 10.1016/j.enpol.2010.02.015.

Demirbas, Ayhan. 2006. "Biodiesel Production via Non-catalytic SCF Method and Biodiesel Fuel Characteristics." *Energy Conversion and Management* 47: 2271–82. doi: 10.1016/j.enconman.2005.11.019.

Demirbas, Ayhan, and Gönenç. Arin. 2002. "An Overview of Biomass Pyrolysis." *Energy Sources* 24: 471–82. doi: 10.1080/00908310252889979.

Demirbas, M. Fatih. 2011. "Biofuels from Algae for Sustainable Development." In Special Issue of "Energy from Algae: Current Status and Future Trends." *Applied Energy* 88, no. 10: 3473–80. doi: 10.1016/j.apenergy.2011.01.059.

Denevan, William M. 1992. "The Pristine Myth: The Landscape of the Americas in 1492." *Annals of the Association of American Geographers*. 82, no. 3: 369–85. doi: 10.1111/j.1467-8306.1992.tb01965.x

Devine-Wright, Patrick. 2007. "Reconsidering Public Attitudes and Public Acceptance of Renewable Energy Technologies: A Critical Review." Working Paper 1.4. School of Environment and Development, University of Manchester, Manchester, UK. http://geography.exeter.ac.uk/beyond_nimbyism/deliverables/bn_wp1_4.pdf.

di Giacomo, Gabriele, and Luca Taglieri. 2009. "Renewable Energy Benefits with Conversion of Woody Residues to Pellets." *Energy* 34: 724–31. doi: 10.1016/j. energy.2008.08.010.

di Lucia, Lorenzo. 2010. "External Governance and the EU Policy for Sustainable Biofuels, the Case of Mozambique." *Energy Policy* 38, no. 11: 7395–403. doi: 10.1016/j.enpol.2010.08.015.

di Pardo, Joseph. 2000. "Outlook for Biomass Ethanol Production and Demand." Energy Information Agency, Washington, DC. https://www.agmrc.org/media/cms /biomass_E6EE9065FD69D.pdf.

Dias de Oliveira, Marcelo, Burton Vaughan, and Edward Rykiel. 2005. "Ethanol as Fuel: Energy, Carbon Dioxide Balances, and Ecological Footprint." *Bioscience* 55: 593–602. doi: 10.1641/0006-3568(2005)055[0593:EAFECD]2.0.CO;2.

Dloughy, Jennifer A. 2021. "Biden EPA Backs Limiting Biofuel Quota Waivers for Refineries." *Bloomberg Green*. 22 February 2021. https://www.bloomberg .com/news/articles/2021-02-22/biden-epa-backs-limiting-biofuel-quota-waivers -for-refineries.

Dogwood Alliance. 2013. "Biomass Campaign Platform." http://docs.nrdc.org/ energy/files/ene_13041702a.pdf.

Duffield, James A., and Keith Collins. 2006. "Evolution of Renewable Energy Policy." *Choices* 21, no. 1: 15–19. http://www.choicesmagazine.org/2006-1/bio-fuels/2006-1-02.htm.

Dunlap, Riley E., and Aaron M. McCright. 2015. "Challenging Climate Change: The Denial Countermovement." In *Climate Change and Society: Sociological Perspectives*, edited by Riley E Dunlap and Robert J. Brulle, 300–32. Oxford: Oxford University Press.

Dwivedi, Puneet, and Janaki R. R. Alavalapati. 2009. "Stakeholders' Perceptions on Forest Biomass-Based Bioenergy Development in the Southern U.S." *Energy Policy* 37: 1999–2007. doi: 10.1016/j.enpol.2009.02.004

Dwivedi, Puneet, Madhu Khannab, Ajay Sharmac, and Andres Susaetad. 2016. "Efficacy of Carbon and Bioenergy Markets in Mitigating Carbon Emissions on Reforested Lands: A Case Study from Southern United States." *Forest Policy and Economics* 67: 1–9. doi: 10.1016/j.forpol.2016.03.002

Dwivedi, Puneet, Robert Bailis, Todd G. Bush, and Marian Marinescu. 2011. "Quantifying GWI of Wood Pellet Production in the Southern United States and its Subsequent Utilization for Electricity Production in The Netherlands/Florida." *Bioenergy Resources* 4: 180–192. doi: 10.1007/s12155-010-9111-5.

Dyer, Janice F., and Conner Bailey. 2008. "A Place to Call Home: Cultural Understandings of Heir Property among Rural African Americans." *Rural Sociology* 73, no. 3: 317–38. doi: 10.1526/003601108785766598.

Dyer, Janice F., Brajesh Singh, and Conner Bailey. 2013. "Differing Perspectives on Biofuels: Analysis of National, Regional, and State Newspaper Coverage." *Journal of Rural Social Sciences* 28, no. 1: 106–33.

Eaton, Weston M., Stephen P. Gasteyer, and Lawrence Busch. 2014. "Bioenergy Futures: Framing Sociotechnical Imaginaries in Local Places." *Rural Sociology* 79, no. 2: 227–56. doi: 10.1111/ruso.12027.

EESI (Environmental and Energy Study Institute). 2006. "2005 Year in Review, U.S. Biomass Policy." Environmental and Energy Study Institute, January 4, 2006. https://www.renewableenergyworld.com/2006/01/04/2005-year-in-review-u -s-biomass-energy-policy-41189/#gref.

Eigenbrode, Sanford D., Michael O'Rourke, J. D. Wulfhorst, David M. Althoff, Caren S. Goldberg, Kaylani Merrill, Wayde Morse, Max Nielsen-Pincus, Jennifer Stephens, Leigh Winowiecki, and Nilsa A. Bosque-Pérez. 2007. "Employing Philosophical Dialogue in Collaborative Science." *Bioscience* 57, no. 1: 55–64. doi: 10.1641/B570109.

Eisler, Matthew. 2012. "Science, Silver Buckshot, and 'All of the Above': Negotiating Energy R&D Policy at ARPA-E Energy Summit 2012." *Science Progress.* 2 April 2012. http://scienceprogress.org/2012/04/science-silver-buckshot-and-%E2%80 %9Call-of-the-above%E2%80%9D/.

Engels, Franziska, and Anna Verena Münch. 2015. "The Micro Smart Grid a Materialized Imaginary within the German Energy Transition." *Energy Research and Social Science* 9: 35–42. doi: 10.1016/j.erss.2015.08.024.

Evans, Jason M., and Matthew J. Cohen. 2009. "Regional Water Resource Implications of Bioethanol Production in the Southeastern United States." *Global Change Biology* 15: 2261–73. doi: 10.1111/j.1365-2486.2009.01868.x.

Fairclough, Norman. 2010. *Critical Discourse Analysis: The Critical Study of Language,* 2nd edition. London: Pearson.

Fairhead, James, Melissa Leach, and Ian Scoones. 2012. "Green Grabbing: A New Appropriation of Nature?" *Journal of Peasant Studies* 39: 237–61. doi: 10.1080/03066150.2012.671770.

Falzon, Mark Anthony. 2009. *Multi-sited Ethnography: Theory, Praxis and Locality in Contemporary Research.* Farnham: Ashegate.

FAO. 2004. "Unified Bioenergy Terminology." Food and Agriculture Organization of the United Nations - Forestry Department, Wood Energy Programme. http://www.fao.org/3/j4504e/j4504e00.htm.

Fargione, Joseph, Jason Hill, David Tilman, Stephene Polasky, and Peter Hawthorne. 2008. "Land Clearing and the Biofuel Carbon Debt." *Science* 319: 1235–38. doi: 10.1126/science.1152747.

Fast, Stewart. 2009. "The Biofuels Debate: Searching for the Role of Environmental Justice in Environmental Discourse." *Environments* 37, no. 1: 83–100.

Fehrenbacher, Katie. 2015. "A Biofuel Dream Gone Bad." *Fortune Magazine*, December 15, 2015 issue. http://fortune.com/kior-vinod-khosla-clean-tech/.

Felt, Ulrike. 2015. "Keeping Technologies Out: Sociotechnical Imaginaries and the Formation of Austria's Technopolitical Identity." In *Dreamscapes of Modernity: Sociotechnical Imaginaries and the Fabrication of Power*, edited by Sheila Jasanoff and Sang-Hyun Kim, 103–25. Chicago: University of Chicago Press.

Fenton, Ed. 2009. "Snake Oil or Climate Cure: The Effect of Public Funding on European Bioenergy." Bioenergy and Forests Briefing Note 02. FERN. http://www.fern.org/sites/fern.org/files/snake%20oil%20or%20climate%20cure_0.pdf.

FERN, BirdLife Europe/Royal Society for the Protection of Birds [RSPB], ClientEarth, European Environmental Bureau, Greenpeace (2012) '"EU Joint NGO Briefing: Sustainability Issues for Solid Biomass in Electricity, Heating, and Cooling." http://www.fern.org/sustainabilityissuesforsolidbiomass.

Fernández, Fruela, and Jonathon Evans, editors. 2018. *The Routledge Handbook of Translation and Politics.* London: Routledge.

Feyerabend, Paul. 1962. "Explanation, Reduction and Empiricism." In *Scientific Explanation, Space, and Time,* edited by Herbert Feigl and Grover Maxwell, 28–97. Minnesota Studies in the Philosophy of Science, Volume III. Minneapolis, MN: University of Minneapolis Press.

Fields, Joe. 2018. "The Future of Energy: Top 100 Influencers, Brands and Publications." *Onalytica*, published online February 7, 2018. https://onalytica.com/blog/posts/future-energy-top-100-influencers-brands-publications/.

Finucane, Melissa. L. 2002. "Mad Cows, Mad Corn, and Mad Money: Applying What We Know about the Perceived Risk of Technologies to the Perceived Risk of Securities." *The Journal of Psychology and Financial Markets* 3, no 4: 236–43. doi: 10.1207/S15327760JPFM0304_05.

Fletcher Jr., Robert J., Bruce A. Robertson, Jason Evans, Patrick J. Doran, Janaki R. R. Alavalapati, and Douglas W. Schemske. 2011. "Biodiversity Conservation in the Era of Biofuels: Risks and Opportunities." *Frontiers in Ecology & the Environment* 9, no. 3: 161. doi: 10.1890/090091.

Ford, James D., Tristan Pearce, Graham McDowell, Lea Berrang-Ford, Jesse S. Sayles, and Ella Belfer. 2018. "Vulnerability and Its Discontents: The Past, Present, and Future of Climate Change Vulnerability Research." *Climatic Change* 151: 189–203. doi: 10.1007/s10584-018-2304-1.

Fox, Thomas R., Eric J. Jokela, and H. Lee Allen. 2004. "The Evolution of Pine Plantation Silviculture in the Southern United States." In *Southern Forest Science: Past, Present, and Future*, H. Michael Rauscher, Kurt Johnsen, editors, 63–82. Gen. Tech. Rep. SRS–75. Asheville, NC: U.S. Department of Agriculture, Forest Service, Southern Research Station.

Franquesa, Juame. 2018. *Power Struggles: Dignity, Value, and the Renewable Energy Frontier in Spain* (New Anthropologies of Europe). Bloomington, IN: Indiana University Press.

Fri, Robert. 2010. "Energy's Silver Buckshot." PBS online. https://www.pbs.org/wnet/need-to-know/opinion/energys-silver-buckshot/5333/.

Gal, Susan. 2015. "Politics of Translation." *Annual Review of Anthropology* 44: 225–40. doi: 10.1146/annurev-anthro-102214-013806.

Gans, Herbert J., 1982. "The Participant Observer as a Human Being: Observations on the Personal Aspects of Fieldwork." In *Field Research: A Sourcebook and Field Manual*, edited by Robert G. Burgess, 53–61. London: George Allen and Unwin, Ltd.

Gasparatos, Alexandros, Per Stromberga, and Kazuhiko Takeuchib. 2011. "Biofuels, Ecosystem Services and Human Wellbeing: Putting Biofuels in the Ecosystem Services Narrative." *Agriculture, Ecosystems and Environment* 142: 111–28. doi: 10.1016/j.agee.2011.04.020.

Genus, Audley, Marfuga Iskandarova, Gary Goggins, Frances Fahy, and Senja Laasko. 2021. "Alternative Energy Imaginaries: Implications for Energy Research, Policy Integration and the Transformation of Energy Systems." *Energy Research & Social Science* 73: 1–9. doi: 10.1016/j.erss.2020.101898.

George, Alexander. 2012. "A 'Silver Buckshot' Will Mend Our Fuelish Ways." *Wired* 31 January 2012. Available online at: http://www.wired.com/autopia/2012/01/future-cars/.

Gerhart, Ann. 2010. "Obama Wants It Clear: 'No Silver Bullet'." *Washington Post*, 8 September 2010. http://www.washingtonpost.com/wp-dyn/content/article/2010/09/08/AR2010090805079.html.

Giampietro, Mario, Sergioa Ulgiati, and David Pimentel. 1997. "Feasibility of Large-scale Biofuel Production." *BioScience* 47: 587–600. doi: 10.2307/1313165.

Gibbs, Holly K., Matt Johnston, Jonathon A. Foley, Traey Halloway, Chad Monfreda, Navin Ramankutty, and David Zacks. 2008. "Carbon Payback Times for Crop-Based Biofuel Expansion in the Tropics: The Effects of Changing Yield and Technology." *Environmental Resource Letters* 3: 1–10. doi: 10.1088/1748-9326/3/3/034001.

Gibson, Lisa. 2010. "Georgia Biomass Will Ship Pellets to Europe." *Biomass Power and Thermal*. http://biomassmagazine.com/articles/3546/georgia-biomass-will-ship-pellets-to-europe.

Gielecki, Mark, Fred Mayes, and Lawrence Prete. 2001. "Incentives, Mandates and Government Programs for Promoting Renewable Energy." In *Renewable Energy 2000: Issues and Trends*, edited by U.S. Department of Energy. Energy Information Administration, 1–17. http://www.eia.doe.gov/cneaf/solar.renewables/rea_issues/incent.html.

Gilbert, Jess, Gwen Sharp, and M. Sindy Felin. 2002. "The Loss and Persistence of Black-Owned Farms and Farmland: A Review of the Research Literature And Its Implications." *Southern Rural Sociology* 18(2): 1–30. https://egrove.olemiss.edu/jrss/vol18/iss2/1.

Gillon, S. 2010. "Field of Dreams: Negotiating an Ethanol Agenda in the Midwest United States." *Journal of Peasant Studies* 37, no. 4: 723–48. doi: 10.1080/03066150.2010.512456.

Ginther, M. Seth. 2011. "The Case for Importing Sustainable Industrial Wood Pellets from the U.S." *Pellet Mill Magazine* Fall 2011. https://issuu.com/bbiinternational/docs/fall11_pmm.

Glaser, Aviva, and Patty Glick. 2012. *Growing Risk: Addressing the Invasive Potential of Bioenergy Feedstocks*. Washington, DC: National Wildlife Federation.

Global CSS [Carbon Capture and Storage] Institute. 2019. "Global Status of CCS 2019: Targeting Climate Change." https://www.globalccsinstitute.com/resources/global-status-report/.

Gluck, Carol, and Anna Lowenhaupt Tsing, eds. 2009. *Words in Motion: Toward a Global Lexicon*. Durham, NC: Duke University Press.

Gnansounou, Edgard, Arnaud Dauriat, and Charles E. Wyman. 2005. "Refining Sweet Sorghum to Ethanol And Sugar: Economic Trade-Offs in the Context of North China." *Bioresource Technology* 96: 985–1002. doi: 10.1016/j.biortech.2004.09.015.

Goluboff, Sascha. 2011. "Making African American Homeplaces in Rural Virginia." *Ethos* 39, no. 3: 368–94. doi: 10.1111/j.1548-1352.2011.01198.x.

González-García, Sara, M. Theresa Moreira, and Gumersindo Feijoo. 2010. "Environmental Performance of Lignocellulosic Bioethanol Production from Alfalfa Stems." *Biofuels, Bioproducts & Biorefining* 4, no. 2: 118–31. doi: 10.1002/bbb.204.

Goodwin, Gary C. 1977. "Cherokees in Transition: A Study of Changing Culture and Environment Prior to 1775." Research Paper No. 181, Department of Geography, University of Chicago.

Gordon, Jason S., Alan Barton, and Keenan Adams. 2013. "An Exploration of African American Forest Landowners in Mississippi." *Rural Sociology* 78, no. 4: 473–97. doi: 10.1111/ruso.12014.

Gottlieb, Peter. 1991. "Rethinking the Great Migration: A Perspective from Pittsburgh." In *The Great Migration in Historical Perspective: New Dimensions of Race, Class, and Gender*, edited by J.W. Trotter, Jr., 68–82. Bloomington, IN: Indiana University Press.

Goyke, Noah, Puneet Dwivedi, Sarah Hitchner, John Schelhas, and Marc Thomas. 2019. "Exploring Diversity in Forest Management Outlooks of Southern African American Family Forest Landowners for Ensuring Sustainability of Forestry Resources." *Human Ecology* 47: 264–274. doi: 10.1007/s10745-019-0068-5.

Graham, Robin Lambert, Richard Nelson, John Sheehan, Robert D. Perlack, and Lynn L. Wright. 2007. "Current and Potential U.S. Corn Stover Supplies." *Agronomy Journal* 99: 1–11. doi: 10.2134/agronj2005.0222.

Green Car Congress. 2019. "LanzaTech Moving Forward on Scale-up of Sustainable Aviation Fuels in US and Japan." *Green Car Congress.* November 21, 2019. https://www.greencarcongress.com/2019/11/201091121-lanzatech.html.

Growth Energy. 2021. "American Ethanol Performance Team." https://americaneth anolracing.com/performance-team/.

Gunderson, Paul. D. 2008. "Biofuels and North American Agriculture – Implications for the Health and Safety of North American Producers." *Journal of Agromedicine* 13, no. 4: 219–24. doi: 10.1080/10599240802454569.

Guo, Mingyang, Qianlai Zhuang, Zeli Tan, Narasinha Shurpali, Sari Juutinen, Pirkko Kortelainen, and Pertti J Martikainen. 2020. "Rising Methane Emissions from Boreal Lakes Due to Increasing Ice-Free Days." *Environmental Research Letters* 15, no. 6: 064008. doi: 10.1088/1748-9326/ab8254.

Guo, Zhihua, Caiyun Sun, and Donald L. Grebner. 2007. "Utilization of Forest Derived Biomass for Energy Production in the U.S.A.: Status, Challenges, And Public Policies." *International Forestry Review* 9, no. 3: 748–58. doi: 10.1505/ ifor.9.3.748.

Halvorson, Kathleen E., Justin R. Barnes, and Barry D. Solomon. 2009. "Upper Midwestern USA Ethanol Potential from Cellulosic Materials." *Society and Natural Resources* 22, no. 10: 931–38. doi: 10.1080/08941920902755382.

Hanks, William F. 2014. "The Space of Translation." *HAU: Journal of Ethnographic Theory* 4, no. 2: 17–39. doi: 10.14318/hau4.2.002.

Haraway, Donna. 1991. "A Cyborg Manifesto: Science, Technology, and Socialist Feminism in the Late Twentieth Century." In *Simians, Cyborgs, and Women: The Reinvention of Nature,* edited by Donna Haraway, 149–82. New York: Routledge.

Harsanyi, David. 2007. "Ethanol is Politicians' Snake Oil." *Denver Post.* 15 February 2007. http://www.denverpost.com/news/ci_5229294.

Hart, John Fraser. 1980. "Land Use Change in a Piedmont County." *Annals of the Association of American Geographers* 70, no. 4: 492–527. doi: 10.1111/j.1467-8306.1980.tb01329.x.

Heal, Geoffrey, and Bengt Kriström. 2002. "Uncertainty and Climate Change." *Environmental and Resource Economics* 22: 3–39.

Heffron, Raphael, Rory Connor, Penelope Crossley, Vincente López-Ibor Mayor, Kim Talus, and Joseph Tomain. 2021. "The Identification and Impact of Justice Risks to Commercial Risks in the Energy Sector: Post COVID-19 and for the Energy Transition." *Journal of Energy and Natural Resources Law.* Publis doi: 10.1080/02646811.2021.1874148.

Herod, Andrew, and Melissa A. Wright. 2002. "Placing Scale: An Introduction." In *Geographies of Power: Placing Scale,* edited by Andrew Herod and Melissa A. Wright, 1–14. Malden, MA: Blackwell Publishers.

Herrando-Pérez, Salvador, Corey J A Bradshaw, Stephan Lewandowsky, and David R Vieites. 2019. "Statistical Language Backs Conservatism in Climate-Change Assessments." *BioScience* 69, no. 3: 209–19. doi: 10.1093/biosci/biz004.

Hill, Jason, Erik Nelson, David Tilman, Stephen Polasky, and Douglas Tiffany. 2006. "Environmental, Economic, and Energetic Costs and Benefits of Biodiesel and

Ethanol Biofuels." *Proceedings of the National Academy of Sciences* 103, no. 30: 11206–10. doi: 10.1073/pnas.0604600103.

Hill, Jason, Stephen Polasky, Erik Nelson, David Tilman, Hong Huo, Lindsay Ludwig, James Neumann, Haochi Zheng, and Diego Bonta. 2009. "Climate Change and Health Costs of Air Emissions from Biofuels and Gasoline." *Proceedings of the National Academy of Sciences* 106, no. 6: 2077–82. doi: 10.1073/pnas.0812835106.

Hilliard-Clark, Joyce, and Clyde E. Chesney. 1985. "Black Woodland Owners: Profile." *Journal of Forestry* 83, no. 11: 674–79. doi: 10.1093/jof/83.11.674.

Himmelfarb, David, John Schelhas, Sarah Hitchner, Cassandra Johnson Gaither, Katherine Dunbar, and J. Peter Brosius. 2014. "Perceptions of and Attitudes toward Climate Change in the Southeastern United States." In *International Perspectives on Climate Change: Latin America and Beyond,* edited by Walter Leal Filho, Fátima Alves, Sandra Caeiro, and Ulisses M. Azeiteiroed, 287–99. New York: Springer Press.

Hinchee, Maud, William Rottmann, Lauren Mullinax, Chunsheng Zhang, Shujun Chang, Michael Cunningham, Leslie Pearson, and Narender Nehra. 2009. "Short-Rotation Woody Crops for Bioenergy and Biofuels Applications." *In Vitro Cellular and Developmental Biology.*45, no. 6: 619–29. doi: 10.1007/s11627-009-9235-5.

Hirsch, Paul D., William Adams, J. Peter Brosius, Asim Zia, Nino Bariola, and Juan Luis Dammert Bello. 2010. "Acknowledging Conservation Trade-Offs and Embracing Complexity." *Conservation Biology* 25, no. 2: 259–64. doi: 10.1111/j.1523-1739.2010.01608.x

Hitchner, Sarah L. 2010. "Heart of Borneo as a *Jalan Tikus*: Exploring the Links Between Indigenous Rights, Extractive and Exploitative Industries, and Conservation at the World Conservation Congress, 2008." *Conservation & Society* 8, no. 4: 320–30. doi: 10.4103/0972-4923.78148.

Hitchner, Sarah, Puneet Dwivedi, John Schelhas, and Arundhati Jagadish. 2019. "Gatekeepers, Shareholders, and Evangelists: Expanding Communication Networks of African American Forest Landowners in North Carolina." *Society and Natural Resources* 32, no. 7: 751–67. doi: 10.1080/08941920.2018.1560521.

Hitchner, Sarah and John Schelhas. 2012. "Social Acceptability of Biofuels among Small-scale Forest Landowners in the U.S. South." In *Proceedings of the IUFRO 3.08.00 2012 Small-scale Forestry Conference,* September 24–27, 2012. Amherst, MA: University of Massachusetts.

Hitchner, Sarah, John Schelhas, Teppo Hujala, and J. Peter Brosius. 2014. "Public Opinion on Wood-Based Bioenergy." In *Wood Energy in Developed Economies: Resource Management, Economics and Policy,* edited by Francisco Aguilar, 32–74. London: Earthscan.

Hitchner, Sarah, John Schelhas, and J. Peter Brosius. 2016. "Snake Oil, Silver Buckshot, and People Who Hate Us: Metaphors and Conventional Discourses of Wood-Based Bioenergy in the Rural Southeastern U.S." *Human Organization* 73: 204–17. doi: 10.17730/1938-3525-75.3.204.

Hitchner, Sarah, John Schelhas, and J. Peter Brosius. 2017. "'Even Our Dairy Queen Shut Down': Risk and Resilience in Bioenergy Development in Forest-dependent

Communities in the U.S. South." *Economic Anthropology* 4, no. 2: 186–199. doi: 10.1002/sea2.12087.

Hitchner, Sarah, John Schelhas, and Cassandra Johnson Gaither. 2017. "A Privilege and a Challenge": Valuation of Heirs' Property by African American Landowners and Implications for Forest Management in the Southeastern U.S." *Small-Scale Forestry* 16: 395–417. doi: /10.1007/s11842-017-9362-5.

Holifield, Ryan, Jayajit Chakraborty, and Gordon Walker, eds. 2018. *The Routledge Handbook of Environmental Justice*. London: Routledge.

Hornborg, A., J.R. McNeill, and J. Martinex Alier. 2007. *Rethinking Environmental History: World System History and Global Environmental Change*. Walnut Creek, CA: Altamira Press.

Hoskinson, Reed L., Douglas L. Karlen, Stuart J. Birrell, Corey W. Radtke, and Wallace W. Wilhelm. 2007. "Engineering, Nutrient Removal, and Feedstock Conversion Evaluations of Four Corn Stover Harvest Scenarios." *Biomass and Bioenergy* 31: 126–36. doi: 10.1016/j.biombioe.2006.07.006.

Huff, Darrell, and Irving Geis. 1954. *How to Lie with Statistics*. New York: W.W. Norton and Company, Inc.

Huggins, David R., Gregory A. Buyanovky, George H. Wagner, James R. Brown, Roberte G. Darmody, Ted R. Peck, Gary W. Lesoing, Matias B. Vanotti, and Larry G. Bundy. 1998. "Soil Organic C in the Tallgrass Prairie-Derived Region of the Corn Belt: Effects of Long-Term Crop Management." *Soil and Tillage Research* 47: 219–34. doi: 10.1016/S0167-1987(98)00108-1.

Hule, Mike. 2009. *Why We Disagree About Climate Change Understanding Controversy, Inaction, and Opportunity*. Cambridge: Cambridge University Press.

Hurlbut, J. Benjamin. 2015. "Remembering the Future: Science, Law, and the Legacy of Asilomar." In *Dreamscapes of Modernity: Sociotechnical Imaginaries and the Fabrication of Power*, edited by Sheila Jasanoff and Sang-Hyun Kim, 126–51. Chicago: University of Chicago Press.

Hurrell, Andrew and Sandeep Sengupta. 2012. "Emerging powers, North–South Relations and Global Climate Politics." *International Affairs* 88, no. 3: 463–84. doi: 10.1111/j.1468-2346.2012.01084.x.

Idso, Craig, Robert M. Carter, and S. Fred Singer. 2015. *Why Scientists Disagree about Global Warming: The NIPCC Report on Scientific Consensus*. Arlington Heights, IL: Heartland Institute.

IEA. 2001. "Experiences from Coal Industry Restructuring." International Energy Agency, Paris, report IEA/CERT/RE/WP(02)16, September, Paris, 25 pp.

IEA. 2002a. "Socio-Economic Aspects of Bioenergy Systems." International Energy Agency Bioenergy Agreement, Activity 29. www.eihp.hr/task29.html.

IEA. 2002b. "Employment Benefits of Renewable Energies." International Energy Agency Internal Institute of Technology, Atlanta.

Institute for Building Efficiency. 2014. "Silver Buckshot: A Variety of Renewable Energy Technologies to Reach the Net Zero Energy Building Vision." http://www.institutebe.com/Green-Net-Zero-Buildings/renewable-energy-technology.aspx.

IPCC [Intergovernmental Panel on Climate Change]. 2014. *Climate Change 2014: Synthesis Report. Contribution of Working Groups I, II and III to the Fifth*

Assessment Report of the Intergovernmental Panel on Climate Change [Core Writing Team, Rajendra K. Pachauri and Leo A. Meyer (editors)]. Geneva, Switzerland: IPCC. 151 pp.

Jacobs, Jennifer, Jennifer A. Dlouhy, and Mario Parker. 2019. "Trump invokes Taliban in Latest Negotiations on Biofuel Plan." *Bloomberg*, September 19, 2019. https://www.bloomberg.com/news/articles/2019-09-19/trump-invokes-taliban-in-latest-deliberations-over-biofuel-plan.

Janowiak, Maria K., and Christopher R. Webster. 2010. "Promoting Ecological Sustainability in Woody Biomass Harvesting." *Journal of Forestry* 108, no. 1: 16–23. doi: 10.1093/jof/108.1.16.

Jasanoff, Sheila and Sang-Hyun Kim. 2009. "Containing the Atom: Sociotechnical Imaginaries and Nuclear Power in the United States and South Korea." *Minerva* 47: 119–46. doi: 10.1007/s11024-009-9124-4.

Jasanoff, Sheila. 2015a. "Future Imperfect: Science, Technology, and the Imaginations of Modernity." In *Dreamscapes of Modernity: Sociotechnical Imaginaries and the Fabrication of Power*, edited by Sheila Jasanoff and Sang-Hyun Kim, 1–33. Chicago: University of Chicago Press.

Jasanoff, Sheila. 2015b. "Imagined and Invented Worlds." In *Dreamscapes of Modernity: Sociotechnical Imaginaries and the Fabrication of Power*, edited by Sheila Jasanoff and Sang-Hyun Kim, 321–41. Chicago: University of Chicago Press.

Jasanoff, Sheila, and Sang-Hyun Kim. 2013. "Sociotechnical Imaginaries and National Energy Policies." *Science as Culture* 22, no. 2: 189–96. doi: 10.1080/09505431.2013.786990.

Jasanoff, Sheila and Sang-Hyun Kim, eds. 2015. *Dreamscapes of Modernity: Sociotechnical Imaginaries and the Fabrication of Power*. Chicago, IL: The University of Chicago Press.

Jenkins, Kirsten, Darren McCauley, Raphael Heffron, Hannes Stephan, and Robert Rehner. 2016. "Energy Justice: A Conceptual Review." *Energy Research & Social Science* 11: 174–82. doi: 10.1016/j.erss.2015.10.004.

Jenkins, Kirsten E. H., Benjamin K. Sovacool, Niek Mouter, Nick Hacking, Mary-Kate Burns, and Darren McCauley. 2021. "The Methodologies, Geographies, and Technologies of Energy Justice: A Systematic and Comprehensive Review." *Environmental Research Letters* 16, no. 4: 043009. doi: 10.1088/1748-9326/abd78c.

Jeon, Chihyung and Yeonsil Kang. 2019. "Restoring and Re-Restoring the CC: Nature, Technology, and History in Seoul, South Korea." *Environmental History* 24, no. 4: 736–65. doi: S10.1093/envhis/emz032.

Jeuck, James, and Doug Duncan. 2009. *Economics of Harvesting Woody Biomass in North Carolina*. Raleigh, NC: NC State Extension. https://content.ces.ncsu.edu/economics-of-harvesting-woody-biomass-in-north-carolina.

Johnson, Cassandra Y., and Josh McDaniel. 2004. "Turpentine Negro." In *To Love the Wind and the Rain: African Americans and Environmental History*, edited by Dianne D. Glave and Mark Stoll, 50–62. Pittsburgh, PA: University of Pittsburgh Press.

Johnson, Jane M. F., Nancy W. Barbour, and Sharon Lachnicht Weyers. 2007. "Chemical Composition of Crop Biomass Impacts its Decomposition." *Soil Science Society of America Journal* 71: 155–62. doi: 10.2136/sssaj2005.0419.

Johnson, Jane M. F., Mark D. Coleman, Russ Gesch, Abdullah Jaradat, Rob Mitchell, Don Reicosky, and Wallace W. Wilhelm. 2007. "Biomass-bioenergy Crops in the United States: A Changing Paradigm." *The Americas Journal of Plant Science and Biotechnology* 1, no. 1: 1–28.

Johnson, Jane M. F., Don C. Reicosky, Ray R. Allmaras, Dave Archer, and Wallace W. Wilhelm. 2006. "A Matter of Balance: Conservation and Renewable Energy." *Journal of Soil and Water Conservation* 61: 120A–25A.

Johnson Gaither, Cassandra, David Himmelfarb, Sarah Hitchner, John Schelhas, Marshall

Shepherd, and Binita K. C. 2016. "'Where the Sidewalk Ends': Climate Change Mitigation and Transportation in Atlanta's Cascade Community." *City & Society* 28, no. 2: 174–97. doi: 10.1111/ciso.12077

Jones, Gerald Everett. 2018 [1995]. *How to Lie with Charts*, 4th Edition. Santa Monica, CA: LaPuerta.

Jones, Landons Y. 2014. "Loaded Language: The Gun Metaphors That Pervade Our Everyday Slang." *Washington Post*, 18 April 2014. http://www.washingtonpost.com/opinions/loaded-language-the-gun-metaphors-that-pervade-our-everyday-slang/2014/04/18/40c4053c-c3ed-11e3-b574-f8748871856a_story.html.

Jones, Michael D. 2014. "Communicating Climate Change: Are Stories Better Than 'Just the Facts'?" *Policy Studies Journal* 42, no. 4: 644–73. doi: 10.1111/psj.12072.

Jørgensen, Uffe. 2011. "Benefits Versus Risks of Growing Biofuels Crops: The Case of *Miscanthus*." *Current Opinion in Environmental Sustainability* 3, no. 1: 24–30. doi: 10.1016/j.cosust.2010.12.003.

Keifer, Captain T.A. "Ike." 2013. "Twenty-First Century Snake Oil: Why the United States Should Reject Biofuels as Part of a Rational National Security Energy Strategy." WICI Occasional Paper No. 4. Waterloo Institute for Complexity and Innovation. http://wici.ca/new/wp-content/uploads/2013/04/Kiefer-Snake-Oil31.pdf.

Kelly, Stephanie. 2020. "Biden Transition Team Holds Talks with Biofuel Groups." *Reuters*, December 24, 2020. https://www.reuters.com/article/us-usa-biofuels-biden/biden-transition-team-holds-talks-with-biofuel-groups-sources-idUSKBN28Y1EL.

Kempton Willett, James S. Boster, Jennifr A. Hartley. 1996. *Environmental Values in American Culture*. Cambridge: MIT Press.

Khooshie, Ranee, and Lal Panjabi. 1993. "Can International Law Improve the Climate - An Analysis of the United Nations Framework Convention on Climate Change Signed at the Rio Summit in 1992." *North Carolina Journal of International Law* 18, no. 3: 491–549.

Kim, J., and K. M. Hayes. 2006. "Southern Pine as Feedstock for Renewable Fuel Industry: A Feasibility Study of Lignocellulose Ethanol Production in Georgia." Enterprise Innovation Institute, Georgia.

Kim, Sang-Hyun. 2015. "Social Movements and Contested Sociotechnical Imaginaries in South Korea." In *Dreamscapes of Modernity: Sociotechnical Imaginaries and the Fabrication of Power*, edited by Sheila Jasanoff and Sang-Hyun Kim, 162–73. Chicago: University of Chicago Press.

Kirby, Jack Temple. 1987. *Rural Worlds Lost: The American South 1920–1960*. Baton Rouge, LA: Louisiana State University Press.

Kirkels, Arjan F. 2012. "Discursive Shifts in Energy from Biomass: A 30-year European Overview." *Renewable and Sustainable Energy Reviews* 16: 4105–15. doi: 10.1016/j.rser.2012.03.037.

Klein, Naomi. 2014. *This Changes Everything: Capitalism Vs. The Climate*. New York: Simon & Schuster.

Kluger, Jeffrey. 2015. "Senator Throws Snowball! Climate Change Disproven!" *Time Magazine*. February 27, 2015. https://time.com/3725994/inhofe-snowball-climate/.

Koester, Stefan, and Sam Davis. 2018. "Siting of Wood Pellet Production Facilities in Environmental Justice Communities in the Southeastern United States." *Environmental Justice* 11, no. 2: 64–70. doi: 10.1089/env.2017.0025.

Koh, Lian Pin, and David Wilcove. 2008. "Is Oil Palm Agriculture Really Destroying Tropical Agriculture?" *Conservation Letters* 1: 60–64. doi: 10.1111/j.1755-263X.2008.00011.x.

Korte, Cara. 2021. "What Deb Haaland's historic confirmation means to Native Americans." *CBS News*, March 16, 2021. https://www.cbsnews.com/news/deb-haaland-native-american-confirmation-interior-secretary/.

Kszos, Lynne Adams, M. E. Downing, Lynne L. Wright, J. H. Cushman, S. B. McLaughlin, V. R. Tolbert, G. A. Tuskan, and M. E. Walsh. 2001. "Bioenergy Feedstock Development Program Status Report." ORNL/TM-2000/292.

Kuchler, Magdalena. 2014. "Sweet Dreams (are Made of Cellulose): Sociotechnical Imaginaries of Second-Generation Bioenergy in the Global Debate." *Ecological Economics* 107: 431–37. doi: 10.1016/j.ecolecon.2014.09.014

Kuchler, Magdalena, and Björn-Ola Linnér. 2012. "Challenging the Food vs. Fuel Dilemma: Genealogical Analysis of the Biofuel Discourse Pursued by International Organizations." *Food Policy* 37: 581–88. doi: 10.1016/j.foodpol.2012.06.005.

Kuhn, Thomas S. 2012 [1962]. *The Structure of Scientific Revolutions*, 4th Edition. Chicago, IL: University of Chicago Press.

Kulcsar, Laszlo J., Theresa Selfa, and Carmen M. Bain. 2016. "Privileged Access and Rural Vulnerabilities: Examining Social and Environmental Exploitation in Bioenergy Development in the American Midwest." *Journal of Rural Studies* 47: 291–99. doi: 10.1016/j.jrurstud.2016.01.008.

Lacey, Justine, Mark Howden, Christopher Cvitanovic, and R. M. Colvin. 2018. "Understanding and Managing Trust at the Climate Science–Policy Interface." *Nature Climate Change* 8: 22–28. doi: 10.1038/s41558-017-0010-z.

Lane, Jim. 2011. "The Range Fuels Failure." *Biofuels Digest*. http://biofuelsdigest.com/bdigest/2011/12/05/the-range-fuels-failure/.

Lane, Jim. 2012. "Gusher! KiOR Starts Production of US Cellulosic Biofuels at Scale." *Biofuels Digest*, November 9, 2012. https://www.biofuelsdigest.com

/bdigest/2012/11/09/gusher-kior-starts-production-of-us-cellulosic-biofuels-at
-scale/.

Lane, Jim. 2016. "KiOR: The Inside True Story of a Company Gone Wrong. Part 5,
The Collapse." *Biofuels Digest*, November 24, 2016. http://www.biofuelsdigest.com
/bdigest/2016/11/24/kior-the-story-of-a-company-gone-wrong-part-5-the-collapse/.

Lane, Jim. 2017. "Don't Spill That Ethanol! Those Cellulosic Fuels Are Worth $4.33
a Gallon." *Biofuels Digest*, published online June 13, 2017. http://www.biofuels-
digest.com/bdigest/2017/06/13/dont-spill-that-ethanol-those-cellulosic-fuels-are
-worth-4-33-a-gallon/.

Lantiainen, Satu M., Nianfu Song, and Francisco X. Aguilar. 2014. "Public Policy
Promoting Wood Energy in the E.U. and U.S." In *Wood Energy in Developed
Economies: Resource Management, Economics, and Policy,* edited by Francisco
X. Aguilar, 306–27. London: Earthscan.

Lassiter, Luke Eric. 2005. "Collaborative Ethnography and Public Anthropology."
Current Anthropology 46, no. 1: 83–106. doi: 10.1086/425658.

Latour, Bruno. 1993. *We Have Never Been Modern.* Cambridge, MA: Harvard
University Press.

Lattimore, Brenna, Charles Tattersall T. Smith, Brian D. Titus, Inge Stupak,
and Gustaf Engell. 2009. "Environmental Factors in Woodfuel Production:
Opportunities, Risks, And Criteria and Indicators for Sustainable Practices."
Biomass and Bioenergy 33: 1321–42. doi: 10.1016/j.biombioe.2009.06.005.

Laurance, William F. 2007. "Switch to Corn Promotes Amazon Deforestation."
Science 318: 1721. doi: 10.1126/science.318.5857.1721b.

Leavitt, John. 2014. "Words and Worlds: Ethnography and Theories of Translation."
HAU: Journal of Ethnographic Theory 4, no. 2: 193–220. doi: 10.14318/
hau4.2.009.

Lefton, Rebecca, and Daniel J. Weiss. 2010. "Oil Dependence is a Dangerous Habit."
Center for *American Progress*, 13 January 2010. http://www.americanprogress.org
/issues/green/report/2010/01/13/7200/oil-dependence-is-a-dangerous-habit/.

Lemus, Rocky, and Rattan Lal. 2005. "Bioenergy Crops and Carbon Sequestration."
Critical Reviews in Plant Sciences 24: 1–21. doi: 10.1080/07352680590910393.

Levenda, Anthony M., Ingrid Behrsin, and Frencesca Disano. 2021. "Renewable
Energy for Whom? A Global Systematic Review of the Environmental Justice
Implications of Renewable Energy Technologies." *Energy Research & Social
Science* 71: 101837. doi: 10.1016/j.erss.2020.101837.

Levidow, Les, and Theo Papaioannou. 2013. "State Imaginaries of the Public Good:
Shaping UK Innovation Priorities for Bioenergy." *Environmental Science and
Policy* 30: 36–49. doi: 10.1016/j.envsci.2012.10.008.

Lassiter, Luke Erik. 2005. "Collaborative Ethnography and Public Anthropology."
Current Anthropology 46: 83–106. doi: 10.1086/425658.

Levin, Simon A. 1992. "The Problem of Pattern and Scale in Ecology." *Ecology* 73,
no. 6: 1943–67. doi: 10.2307/1941447.

Lewis, Sophie, and Gallant, Allie. 2013. "In Science, the Only Certainty Is
Uncertainty." https://theconversation.com/in-science-the-only-certainty-is-uncer-
tainty-17180.

Li, Tania M. 1996. "Images of Community: Discourse and Strategy in Property Relations." *Development and Change* 27: 501–27. doi: 10.1111/j.1467-7660.1996. tb00601.x.

Linacre, Nicholas A., and Peter K. Ades. 2004. "Estimating Isolation Distances for Genetically Modified Trees in Plantation Forestry." *Ecological Modelling* 179: 247–57. doi: 10.1016/j.ecolmodel.2003.11.011.

Lindsey, Rebecca. 2020. "Climate Change: Atmospheric Carbon Dioxide." NOAA Climate.gov. August 14, 2020. https://www.climate.gov/news-features/under-standing-climate/climate-change-atmospheric-carbon-dioxide.

Long, Mindy. 2013. "Carriers Use 'Silver Buckshot' to Shoot Down Fuel Costs." *Transport Topics* 4032: A1, A6–A7.

Lora, Electo E. S., José C. Escobar Palacio, Mateus H. Rocha, Maria L. G. Renó, Osvaldo J. Venturini, and Oscar Almazán del Olmo. 2011. "Issues to Consider, Existing Tools and Constraints in Biofuels Sustainability Assessments." *Energy* 36: 2097–110. doi: 10.1016/j.energy.2010.06.012.

Lu, Hangyong R., and Ali El Hanandeh. 2017. "Assessment of Bioenergy Production from Mid-Rotation Thinning of Hardwood Plantation: Life Cycle Assessment and Cost Analysis." *Clean Technologies and Environmental Policy* 19: 2021–40. doi: 10.1007/s10098-017-1386-1.

MacCleery, Douglas W. 2011. *American Forests: A History of Resiliency and Recovery*. Durham, NC: Forest History Society.

Macedo, Isias C., Joaquim E. A. Seabra, and João E. A. R. Silva. 2008. "Greenhouse Gasses Emissions in the Production and Use of Ethanol from Sugarcane in Brazil: The 2005/2006 Averages and a Prediction for 2020." *Biomass and Bioenergy* 32: 582–95. doi: 10.1016/j.biombioe.2007.12.006.

Maclin, Edward M., and Juan Luis Dammert Bello. 2010. "Setting the Stage for Biofuels: Policy Texts, Communities of Practice and Institutional Ambiguity at the Fourth World Conservation Congress." *Conservation & Society* 8, no. 4: 312–19. doi: 10.4103/0972-4923.78147.

Magar, Surya Bahadur, Paavo Pelkonen, Liisa Tahvanainen, Ritva Toivonen, and Anne Toppinen. 2011. "Growing Trade of Bioenergy in the EU: Public Acceptability, Policy Harmonization, European Standards and Certification Needs." *Biomass and Bioenergy* 35, no. 8: 3318–27. doi: 10.1016/j. biombioe.2010.10.012.

Mainville, Nicolas. 2011. *Fuelling a BioMess: Why Burning Trees for Energy Will Harm People, the Climate and Forests*. Montreal, Canada: Greenpeace Canada. http://www.greenpeace.org/canada/Global/canada/report/2011/10/ForestBiomess _Eng.pdf

Malone, Elizabeth L. 2009. *Debating Climate Change: Pathways Through Argument to Agreement*. London: Earthscan.

Mann, Linda K., Virginia R. Tolbert, and Janet Cushman. 2002. "Potential Environmental Effects of Corn (*Zea Mays* L.) Stover Removal with Emphasis on SOM and Erosion: A Review." *Agriculture, Ecosystems and Environment* 89: 149–66. doi: 10.1016/S0167-8809(01)00166-9.

Mann, Michael. 2012. *The Hockey Stick and the Climate Wars: Dispatches from the Front Lines.* New York: Columbia University Press.

Mann, Michael E., Raymond S. Bradley, and Malcolm K. Hughes. 1998. "Global-Scale Temperature Patterns and Climate Forcing Over the Past Six Centuries." *Nature* 392: 779–87.

Marcus, George E. 1995. "Ethnography In/Of the World System: The Emergence of Multi-sited Ethnography." *Annual Review of Anthropology* 24: 95–117. doi: 10.1146/annurev.an.24.100195.000523.

Marland, Gregg, Bernhard Schlamadinger, and Paul Leiby. 1997. "Forest/Biomass-based Mitigation Strategies: Does the Timing of Carbon Reductions Matter?" *Critical Reviews of Environmental Science and Technology* 27: S213–26. doi: 10.1080/10643389709388521.

Marris, Emma. 2006. "Sugar Cane and Ethanol: Drink the Best and Drive the Rest." *Nature* 444: 670–72. doi: 10.1038/444670a

Marshall, Elizabeth P. 2007. "Carving out Space for Sustainability in Biofuel Production." *Agricultural and Resource Economics Review* 36, no. 2: 183–96. doi: 10.22004/ag.econ.44696.

Marston, Sallie A. 2000. "The Social Construction of Scale." *Progress in Human Geography* 24, no. 2: 219–42. doi: 10.1191/030913200674086272.

Mayfield, Chyrel A., C. Darwin Foster, Charles Tattersall Smith, Jianbang Gan, and Susan Fox. 2007. "Opportunities, Barriers, and Strategies for Forest Bioenergy and Bio-Based Product Development in the Southern United States." *Biomass and Bioenergy* 31: 631–37. doi: 10.1016/j.biombioe.2007.06.021.

McBride, Allen. 2011. "Biomass Energy Sustainability." *Biomass Energy Data Book.* Oak Ridge, TN: U.S. Department of Energy.

McCormick, Kes, and Niina Kautto. 2013. "The Bioeconomy in Europe: An Overview." *Sustainability* 5: 2589–608. doi: 10.3390/su5062589.

McCurry, John W. 2011. "Pine Power." *Site Selection Magazine.* January 2011. https://siteselection.com/issues/2011/jan/Georgia.cfm.

Macfarlane, Daniel. 2020. "Nature Empowered: Hydraulic Models and the Engineering of Niagara Falls." *Technology & Culture* 61, no. 1: 109–43. doi: 10.1353/tech.2020.0034.

McGuire, Bill D. 2012. *Assessment of the Bioenergy Provisions in the 2008 Farm Bill.* Washington, DC: Association of Fish and Wildlife Agencies.

McLaughlin, Samuel, and Marie Walsh. 1998. "Evaluating Environmental Consequences of Producing Herbaceous Crops for Bioenergy." *Biomass and Bioenergy* 14: 317–24. doi: 10.1016/S0961-9534(97)10066-6.

McLaughlin, Samuel B., Daniel de la Torre Ugarte, C. T. Garten, Lee R. Lynd, Matt A. Sanderson, Virginia R. Tolbert, and D. D. Wolf. 2002. "High-value Renewable Energy from Prairie Grasses." *Environmental Science and Technology* 36: 2122–29. doi: 10.1021/es010963d.

McShane, Thomas, Paul Hirsch, Tran Chi Trung, Alexander Songorwa, Ann Kinzig, Bruno Monteferri, David Mutekanga, et al. 2011. "Hard Choices: Making Trade-offs between Biodiversity Conservation and Human Well-Being." *Biological Conservation* 144:966–72. doi: 10.1016/j.biocon.2010.04.038.

Mendes Souza, Glaucia, Maria Victoria R. Ballester, Carlos Henrique de Brito Cruz, Helena Chum, Bruce Dale, Virginia H. Dale, Erick C.M. Fernandes, et al. 2017. "The Role of Bioenergy in a Climate-Changing World." *Environmental Development* 23: 57–64. doi: 10.1016/j.envdev.2017.02.008.

Miles, Susan, Øydis Ueland, and Lynn J. Frewer. 2005. "Public Attitudes Towards Genetically-modified Food." *British Food Journal* 107, no. 4: 246–62. doi: 10.1108/00070700510589521.

Miller, Clark A., Jason O'Leary, Elisabeth Graffy, Ellen B. Stechel, and Gary Dirks. 2014. "Narrative Futures and the Governance of Energy Transitions." *Futures*. 70: 65–74. doi: 10.1016/j.futures.2014.12.001.

Miller, Scott C. 2006. "Cellulosic Ethanol – Snake Oil for the New Millennium?" *BIOconversion Blog*, 21 November 2006. http://bioconversion.blogspot.com/2006/11/cellulosic-ethanol-snake-oil-for-new.html.

Miranowski, John A. 2007. "Biofuel Incentives and the Energy Title of the 2007 Farm Bill." In *The 2007 Farm Bill and Beyond*, edited by Bruce L. Gardner and Daniel A. Sumner 121–26. Washington, D.C.: American Enterprise Institute. http://www.aei.org/research/farmbill/publications/pageID.1476,projectID.28/default.asp.

Mitchell, Donald. 2008. "A Note on Rising Food Prices." World Bank Policy Research Working Paper 4682, Development Prospects Group. The World Bank, Washington, DC. https://ssrn.com/abstract=1233058.

Mitchell, Thomas W. 2001. "From Reconstruction to Deconstruction: Undermining Black Landownership, Political Independence, and Community Through Partition Sales of Tenancies in Common." *Northwestern University Law Review* 95, no. 2: 505–80. https://scholarship.law.tamu.edu/facscholar/783.

Mitchell, Thomas W. 2014. Reforming Property Law to Address Devastating Land Loss." *Alabama Law Review* 66, no. 1: 1–61. https://scholarship.law.tamu.edu/facscholar/790.

Mitchell, Thomas W. 2005. "Destabilizing the Normalization of Rural Black Land Loss: A Critical Role for Legal Empiricism." *Wisconsin Law Review* 2: 557–615.

Mittlefehldt, Sarah. 2016. "Seeing Forests as Fuel: How Conflicting Narratives Have Shaped Woody Biomass Development in the United States since the 1970s." *Energy Research & Social Science* 14: 13–21. doi: 10/1016/j.erss.2015/12/023.

Mittlefehldt, Sarah. 2018. "Wood Waste and Race: The Industrialization of Biomass Energy Technologies and Environmental Justice." *Technology and Culture* 59, no. 4: 875–98. doi: 10.1353/tech.2018.0089.

Monmonier, Mark. 1991. *How to Lie with Maps*. Chicago IL: University of Chicago Press.

Monroe, Martha C., and Annie Oxarart. 2011. "Woody Biomass Outreach in the Southern United States: A Case Study." *Biomass and Bioenergy* 35: 1465–73. doi: 10.1016/j.biombioe.2010.08.064.

Mooney, Chris. 2013. "The Hockey Stick: The Most Controversial Chart in Science, Explained." *The Atlantic* May 10, 2013. https://www.theatlantic.com/technology/archive/2013/05/the-hockey-stick-the-most-controversial-chart-in-science-explained/275753/.

Nap, Jan-Peter, Peter Metz, Marga Escaler, and Anthony J. Conner. 2003. "The Release of Genetically Modified Crops into the Environment." *The Plant Journal* 33, no: 1–18. doi: 10.1046/j.0960-7412.2003.01602.x.

National Research Council (NRC). 2007. *Water Implications of Biofuels Production in the United States.* Washington, DC: National Academies Press.

Naylor, Rosamond L., Adam J. Liska, Marshall B. Burke, Walter P. Falcon, Joanne C. Gaskell, Scott D. Rozelle and Kenneth G. Cassman. 2007. "The Ripple Effect: Biofuels, Food Security, and the Environment." *Environment: Science and Policy for Sustainable Development* 49, no. 9: 30–43. doi: 10.3200/ENVT.49.9.30-43.

Nazzaro, Robin M. 2005. "Natural Resources: Federal Agencies are Engaged in Various Efforts to Promote the Utilization of Woody Biomass, But Significant Obstacles to Its Use Remain." GAO-05-373 Uses of Woody Biomass. United States Government Accountability Office. 51 pp. https://digital.library.unt.edu/ark:/67531/metadc297617/m1/1/.

Nelson, Richard G. 2002. "Resource Assessment and Removal Analysis for Corn Stover and Wheat Straw in the Eastern and Midwestern United States – Rainfall and Wind-Induced Soil Erosion Methodology." *Biomass and Bioenergy* 22: 349–63. doi: 10.1016/S0961-9534(02)00006-5.

Nepstad, Daniel, Georgia Carvalho, Ana Cristina Barros, Ane Alencar, João Paulo Capobianco, Josh Bishop, Paulo Moutinho, Paul Lefebvre, Urbano Lopes Silva, and Elaine Prins. 2001. "Road Paving, Fire Regime Feedbacks, and the Future of Amazon Forests." *Forest Ecology Management* 154: 395–407. doi: 10.1016/S0378-1127(01)00511-4.

Nettles, Jami, Peter Birks, Eric Sucre, and Robert Bilby. 2015. "Sustainable Production of Bioenergy Feedstock from the Industrial Forest: Potential and Challenges of Operational Scale Implementation." *Current Sustainable/Renewable Energy Reports* 2: 121–27. doi: 10.1007/s40518-015-0042-9.

Newfont, Kathryn. 2012. *Blue Ridge Commons: Environmental Activism and Forest History in Western North Carolina.* Athens, GA: University of Georgia Press.

Newman, David H., and David N. Wear. 1993. "Production Economics of Private Forestry: A Comparison of Industrial and Nonindustrial Forest Owners." *American Journal of Agricultural Economics* 75: 674–84. doi: 10.2307/1243574.

Newsom, Deanna, Volker Bahn, and Benjamin Cashore. 2006. "Does Forest Certification Matter? An Analysis of Operational Changes Required During the Smartwood Certification Process in the United States." *Forest Policy and Economics* 9: 197–208. doi: 10.1016/j.forpol.2005.06.007.

Newsom, Deannna, and Daphne Hewitt. 2001. "The Global Impacts of SmartWood Certification." Final Report. Rainforest Alliance. 39 pp. http://www.rainforest-alliance.org/sites/default/files/publication/pdf/sw_impacts.pdf

Olsson, Per, Carl Folke, and Fikret Berkes. 2004. "Adaptive Management for Building Resilience in Social-ecological Systems." *Environmental Management* 34, no. 1: 75–90. doi: 10.1007/s00267-003-0101-7.

Omer, Abdeen Mustafa. 2013. "Recent Advances in Biomass and Biogas Development and Perspectives: Bringing Together Geographical and Sociological Imaginations." *Scientific Journal of Biological Sciences* 2, no. 9: 190–207.

Oreskes, Naomi and Erik M. Conway. 2010. *Merchants of Doubt: How a Handful of Scientists Obscured the Truth on Issues from Tobacco Smoke to Global Warming.* New York: Bloomsbury Press.

Oswald, Gordon K. A., and S. P. Gogineni. 2012. "Mapping Basal Melt Under the Northern Greenland Ice Sheet." *IEEE Transactions on Geoscience and Remote Sensing* 50, no. 2: 585–92. doi: 10.1109/TGRS.2011.2162072.

Oswalt, Sonia N., and W. Brad Smith, editors. 2014. "U.S. Forest Resource Facts and Historical Trends." FS-1035. USDA Forest Service, Washington, DC.

Oswalt, Sonia N., W. Brad Smith, Patrick D. Miles, and Scott Pugh. 2014." Forest Resources of the United States, 2012: A Technical Document Supporting the Forest Service 2015 Update of the RPA Assessment." General Technical Report WO-91. USDA Forest Service, Washington, DC.

Outka, Uma. 2012. "Environmental Justice Issues in Sustainable Development: Environmental Justice in the Renewable Energy Transition." *Journal of Environmental and Sustainability Law* 19, no. 1: 60.

Outland, Robert B. III. 1996. "Slavery, Work, and the Geography of the North Carolina Naval Stores Industry, 1835–1860." *The Journal of Southern History* 62, no. 1: 27–56. doi: 10.2307/2211205.

Overdevest, Christinea and Mark G. Rickenbach. 2006. "Forest Certification and Institutional Governance: An Empirical Study of Forest Stewardship Council Certificate Holders in the United States." *Forest Policy and Economics* 9: 93–102. doi: 10.1016/j.forpol.2005.03.014.

Overland, Indra, and Benjamin K. Sovacool. 2020. "The Misallocation of Climate Research Funding." *Energy Research and Social Science* 62: 101349. doi: 10.1016 .j.erss.2019.101349.

Owen, W. 2002. "The History of Native Plant Communities in the South." In *Southern Forest Resource Assessment*, edited by David N. Wear and John G. Greis, 47–61. General Technical Report SRS-53. Asheville, NC: U.S. Department of Agriculture Forest Service Southern Research Station.

Palmer, James. 2021. "Putting Forests to Work? Enrolling Vegetal Labor in the Socioecological Fix of Bioenergy Resource Making." *Annals of the American Association of Geographers* 111, no. 1: 141–156. doi: 10.1080/24694452.2020.1749022.

Pätäri, Satu. 2010. "Industry-and Company-level Factors Influencing the Development of the Forest Energy Business—Insights from a Delphi Study." *Technological Forecasting and Social Change* 77, no. 1: 94–109. doi: 10.1016/j. techfore.2009.06.004.

Pate, Ron, Geoff Klise, and Ben Wu. 2011. "Resource Demand Implications for US Algae Biofuels Production Scale-up." *Applied Energy* 88, no. 10: 3377–88. doi: 10.1016/j.apenergy.2011.04.023.

Payne, Brian R. and Donald R. Theoe. 1971. "Black Foresters Needed: A Professional Concern." *Journal of Forestry* 69, no. 5: 295–98. doi: 10.1093/ jof/69.5.295.

Perlack, Robert D., J. W. Ramney, and Lynn L. Wright. 1992. "Environmental Emissions and Socioeconomic Considerations in the Production, Storage, and

Transportation of Biomass Energy Feedstocks." ORNL/TM-12030. Oak Ridge, TN: Oak Ridge National Laboratory. doi: 10.2172/7256087.

Perlack, Robert, Lynn L. Wright, Anthony F. Turhollow, Robin L. Graham, Bryce J. Stokes, and Donald C. Erbach. 2005. "Biomass as a Feedstock for a Bioenergy and Bioproducts Industry: The Technical Feasibility of a Billion-Ton Annual Supply." US Department of Energy and U.S. Department of Agriculture. Oak Ridge, TN: Oak Ridge National Laboratory.

Petrou, Evangelos C., and Costas P. Pappis. 2009. "Biofuels: A Survey on Pros and Cons." *Energy and Fuels* 23: 1055–66. doi: 10.1021/ef800806g.

Pfeffer, Max J., John Schelhas, and Leyla Ann Day. 2001. "Forest Conservation, Value Conflict, and Interest Formation in a Honduran National Park." *Rural Sociology* 66, no. 3: 382–402. doi: 10.1111/j.1549-0831.2001.tb00073.x.

Phadke, Roopali. 2018. "Green Energy Futures: Responsible Mining on Minnesota's Iron Range." *Energy Research and Social Science* 35: 163–73. doi: 10.1016/j.erss.2017.10.036.

Phillips, Ari. 2015. "American Companies Are Shipping Millions of Trees to Europe, and It's a Renewable Energy Nightmare." *Climate Progress.* https://thinkprogress.org/american-companies-are-shipping-millions-of-trees-to-europe-and-its-a-renewable-energy-nightmare-ef1fdcca1cb4#.gy0p0alup.

Pimentel, David, and Tad W. Patzek. 2005. "Ethanol Production using Corn, Switchgrass, and Wood; Biodiesel Production using Soybean and Sunflower." *Natural Resources Research* 14: 65–76. doi: 10.1007/978-1-4020-8654-0_15.

Plate, Richard R., Martha C. Monroe, and Annie Oxarart. 2010. "Public Perceptions of Using Woody Biomass as a Renewable Energy Source." *Journal of Extension* 48, no. 3: 1–15. Article Number 3FEA7.

Polanyi, Michael. 1958. *Personal Knowledge: Towards a Post-critical Philosophy.* London: Routledge.

Portz, Tim. 2012. "Biomass' Own Silver Buckshot." *Biomass Magazine.* 22 August 2012. http://biomassmagazine.com/articles/7967/biomassundefined-own-silver-buckshot.

Portz, Tim. 2014. "Considering the Counterfactuals." *Biomass Magazine.* August 05, 2014. http://biomassmagazine.com/blog/article/2014/08/considering-the-counterfactuals.

Poudyal, Neelam C., Omkar Joshi, Adam M. Taylor, and Donald G. Hodges. 2017. "Prospects of Wood-Based Energy Alternatives in Revitalizing the Economy Impacted by Decline in the Pulp and Paper Industry." *Forest Products Journal* 67, no. 7/8: 427–34. doi: 10.13073/FPJ-D-17-00004.

Povenelli, Elizabeth. 2001. "Radical Worlds: The Anthropology of Incommensurability and Inconceivability." *Annual Review of Anthropology* 30: 319–34. doi: 10.1146/annurev.anthro.30.1.319.

Practical Law Environment. 2020. "COVID-19: UNFCCC COP 26 Climate Summit and CBD Biodiversity Summit Postponed Until 2021." April 2, 2020. Thompson Reuters. https://uk.practicallaw.thomsonreuters.com/w-024-8135?transitionType=Default&contextData=(sc.Default)&firstPage=true.

Prandi, Pierre, Benoit Meyssignac, Michaël Ablain, Giorgio Spada, Aurélien Ribes, and Jérôme Benveniste. 2021. "Local Sea Level Trends, Accelerations and Uncertainties over 1993–2019." *Scientific Data* 8, no. 1. doi: 10.1038/s41597-020-00786-7.

Pritchard, Sara B. 2011. *Confluence: The Nature of Technology and the Remaking of the Rhone*. Cambridge, MA: Harvard University Press.

Pritchard, Sara B. 2013. "Joining Environmental History with Science and Technology Studies: Promises, Challenges, and Contributions." In *New Natures: Joining Environmental History with Science and Technology Studies*, edited by Dolly Jørgensen, Finn Arne Jørgensen, and Sara B. Pritchard, 1–17. Pittsburgh, PA: University of Pittsburgh Press.

Pritzker, Penny, Kathryn Sullivan, Russell Callender, Richard Edwing, William V. Sweet. Robert E. Christopher, P. Weaver, et al. 2017. "Global and Regional Sea Level Rise Scenarios for the United States." NOAA Technical Report NOS CO-OPS 083. U.S. Department of Commerce, Silver Springs, MD.

Putnam, D. H., J. T. Budin, L. A. Field, and W. M. Breene. 1993. "*Camelina*: A Promising Low-input Oilseed." In *New Crops,* edited by Jules Janick J. and James E. Simon, 314–22. New York: Wiley.

Quaranda, Scot. 2008. *Don't Log the Forests for Fuel: A Position Paper on the Potential Environmental and Economic Impacts of the Cellulosic Ethanol Industry in the Southern United States*. Asheville, NC: Dogwood Alliance.

Quinn, Naomi. 2005. *Finding Culture in Talk: A Collection of Methods*. New York: Palgrave MacMillan.

Quinn, Naomi, and D. Holland. 1987. "Culture and Cognition." In *Cultural Models in Language and Thought*, edited by D. Holland and Naomi Quinn, 3–40. Cambridge, MA: MIT Press.

Radics, Robert I., Sudipta Dasmohapatra, and Stephen S. Kelley. 2016. "Public Perception of Bioenergy in North Carolina and Tennessee." *Energy, Sustainability, and Society* 6, no. 17: 1–11. doi: 10.1186/s13705-016-0081-0.

Reijinders, L. 2006. "Conditions for the Sustainability of Biomass Based Fuel Use." *Energy Policy* 34: 863–76. doi: 10.1016/j.enpol.2004.09.001.

Reilly, John, Peter H. Stone, Chris E. Forest, Mort D. Webster, Henry D. Jacoby, and Ronald G. Prinn. 2001. "Uncertainty and Climate Change Assessments." *Science* 293, no. 5529: 430–33. doi: 10.1126/science.1062001.

Reitze, Arnold W. 2009. "Biofuels: Snake Oil for the Twenty-first Century." *Oregon Law Review* 87: 1183. https://ssrn.com/abstract=1554647.

Righelato, Renton and Dominick V. Spracklen. 2007. "Environment-Carbon Mitigation by Biofuels or by Saving and Restoring Forests?" *Science* 317, no. 5820: 902. doi: 10.1126/science.1141361.

Rohracher, Harald, Thomas Späth, Philipp and Florian Faber. 2004. "Improving the Public Perception of Bioenergy in the EU." Report to the Directorate-General for Energy and Transport, European Commission, http://ec.europa.eu/energy/res/sectors/doc/bioenergy/bioenergy_perception.pdf.

Roncoli, Carla, Todd Crane, and Ben Orlove. 2009. "Fielding climate change in cultural anthropology. In *Anthropology and Climate Change*, edited by Susan A. Crate and Mark Nuttall, editors, 87–115. Walnut Creek, CA: Left Coast Press,.

Rossi, Andrea, and Yianna Lambrou. 2008. *Gender and Equity Issues in Liquid Biofuels Production: Minimizing the Risks to Maximize the Opportunities*. Rome: FAO. ftp://ftp.fao.org/docrep/fao/010/ai503e/ai503e00.pdf.

RSPB, Friends of the Earth, and Greenpeace. 2012. "Dirtier than Coal? Why Government Plans to Subsidise Burning Trees are Bad News for the Planet." http://www.rspb.org.uk/Images/biomass_report_tcm9-326672.pdf.

Rudel, Thomas. 2001. "Did a Green Revolution Reforest the American South?" In *Agricultural Technologies and Tropical Deforestation*, edited by Arild Angelsen and David Kaimowitz, 33–54. New York and London: CABI Press.

Russell, Edmund, James Allison, Thomas Finger, John K. Brown, Brian Balogh, and W. Bernard Carlson. 2011. "The Nature of Power: Synthesizing the History of Technology and Environmental History." *Technology and Culture* 52, no. 2: 246–59. doi: 10.1353/tech.2011.0071.

Sanderson, Matt A., Roderick Reed, Samuel McLaughlin, Stan Duane Wullschleger, Bob V. Conger, David Parrish, Dale Wolf, Charles Taliaferro, Andrew Hopkins, William Ocumpaugh, Mark Hussey, James Read, and Charles Tischler. 1996. "Switchgrass as a Sustainable Bioenergy Crop." *Bioresource Technology* 56: 83–93. doi: 10.1016/0960-8524(95)00176-X.

Schelhas, John. 2002. "Race, Ethnicity, and Natural Resources in the United States: A Review." *Natural Resources Journal* 42, no. 4: 723–63.

Schelhas, John, and Sarah Hitchner. 2020. "Integrating Research and Outreach for Environmental Justice: African American Landownership and Forestry in the U.S. South." *Annals of Anthropological Practice* 77. No. 1: 47–64. doi: 10.1111/napa.12133

Schelhas, John, Sarah Hitchner, and J. Peter Brosius. 2017. "Envisioning and Implementing Sustainable Bioenergy Systems in the U.S. South." In *Handbook of Sustainability and Social Science Research*, edited by Walter Leal Filho, Robert Marans, and John Callewaert, 301–314. World Sustainability Series. Cham, Switzerland: Springer.

Schelhas, John, Sarah Hitchner, and J. Peter Brosius. 2018. "Envisioning and Implementing Wood-based Bioenergy Systems in the Southern United States: Imaginaries in Everyday Talk." *Energy and Social Science* 35: 182–92. doi: 10.1016/j.erss.2017.10.042.

Schelhas, John, Sarah Hitchner, and Puneet Dwivedi. 2018. "Strategies for Successful Engagement of African American Landowners in Forestry." *Journal of Forestry* 116, no. 6: 581–88. doi: 10.1093/jofore/fvy044.

Schelhas, John, Sarah Hitchner, and Cassandra Johnson. 2012. "Social Vulnerability and Environmental Change along Urban-Rural Interfaces." In *Urban–Rural Interfaces: Linking People and Nature*, edited by David N. Laband, B. Graeme Lockaby, and Wayne Zipperer, 171–85. Madison, WI: American Society of Agronomy, Soil Science Society of America, Crop Science Society of America.

Schelhas, John, Sarah Hitchner, Cassandra Johnson Gaither, Rory Fraser, Viniece Jennings, and Amadou Diop. 2017a. "Engaging African American Landowners in Sustainable Forest Management." *Journal of Forestry* 115, no. 1: 26–33. doi: 10.5849/jof.15-116.

Schelhas, John, Sarah Hitchner, and Alan McGregor. 2019. "The Sustainable Forestry and African American Land Retention Program." In *Heirs' Property and Land Fractionation: Fostering Stable Ownership to Prevent Land Loss and Abandonment,* edited by Cassandra J. Gaither, Ann Carpenter, Tracy Lloyd McCurty, and Sarah Toering, 20–8. e-Gen. Tech. Rep. SRS-244. Asheville, NC: U.S. Department of Agriculture Forest Service, Southern Research Station. 105 p.

Schelhas, John, and James P. Lassoie. 2001. "Learning Conservation and Sustainable Development: An Interdisciplinary Approach." *Journal of Natural Resource and Life Sciences Education* 30: 111–19.

Schelhas, John, and Max J. Pfeffer. 2005. "Forest Values of National Park Neighbors in Costa Rica." *Human Organization* 64, no. 4: 385–97.

Schelhas, John, and Max J. Pfeffer. 2008. *Saving Forests, Protecting People? Environmental Conservation in Central America.* (Globalization and Environment Series.) Walnut Creek, CA: AltaMira Press.

Schelhas, John, and Max J. Pfeffer. 2009. "When Global Conservation Meets Local Livelihoods: Policy and Management Lessons." *Conservation Letters* 2, no. 6: 278–85. doi: 10.1111/j.1755-263X.2009.00079.x.

Schelhas, John, Robert Zabawa, and J. J. Molnar. 2003. "New Opportunities for Social Research on Forest Landowners in the South." *Southern Rural Sociology* 19, no. 2: 60–69.

Schelhas, John, Yaoqi Zhang, Robert Zabawa, and Bin Zheng. 2012. "Exploring Family Forest Landowner Diversity: Place, Race, and Gender in Alabama." *International Journal of Social Forestry* 5, no. 1: 1–21.

Scheurich, Greg. 2010. "Ethanol: The Latest Incarnation of Snake Oil." *Energy Tribune*, 18 June 2010. http://oilprice.com/Alternative-Energy/Biofuels/Ethanol-The-Latest-Incarnation-Of-Snake-Oil.html.

Schill, Suzanne Retka. 2014. "LanzaTech to Move Headquarters: Soperton Retrofit Ongoing." April 8, 2014. *Ethanol Producer Magazine.* http://ethanolproducer.com/articles/10934/lanzatech-to-move-headquarters-soperton-retrofit-ongoing

Schlossberg, Josh. 2013. "Biomass Industry Reveals Plans to Turn U.S. into European Resource Colony." *The Biomass Monitor.* August 4, 2013. http://www.energyjustice.net/content/biomass-industry-reveals-plans-turn-us-european-resource-colony

Schmer, Marty R., Kenneth P. Vogel, Rob B. Mitchell, and Richard K. Perrin. 2008. "Net Energy of Cellulosic Ethanol from Switchgrass." *Proceedings of the National Academy of Sciences* 105: 464–69. doi: 10.1073/pnas.0704767105.

Schreurs, Miranda A. 2012. "20th Anniversary of the Rio Summit: Taking a Look Back and at the Road Ahead." *GAIA - Ecological Perspectives for Science and Society* 21, no. 1: 13–16. doi: 10.14512/gaia.21.1.6.

Schwom Rachel L., Aaron M. McCright, and Steven R. Brechin with Riley E. Dunlap, Sandra T. Marquart-Pyatt, and Lawrence Hamilton. 2015. "Public Opinion on Climate Change." In *Climate Change and Society: Sociological Perspectives*, edited by Riley E Dunlap and Robert J. Brulle, 269–99. Oxford: Oxford University Press.

Scott, Deborah, Sarah Hitchner, Edward M. Maclin, and Juan Luis Dammert Bello. 2014. "Fuel for the Fire: Biofuels and the Problem of Translation at the Tenth

Conference of the Parties to the Convention on Biological Diversity." *Global Environmental Politics* 14, no. 3: 84–101. doi: 10.1162/GLEP_a_00240.

Scott, James C. 1985. *Weapons of the Weak: Everyday Forms of Peasant Resistance.* New Haven, CT: Yale University Press.

Scott, James C. 1998. *Seeing Like a State: How Certain Schemes for Improve the Human Condition Have Failed.* New Haven, CT: Yale University Press.

Searchinger, Timothy, Ralph Heimlich, R. A. Houghton, Fengxia X. Dong, Amani Elobeid, Jacinto Fabiosa, Simla Tokgoz, Dermot Hayes, and Tun-Hsiang Yu. 2008. "Use of US Croplands for Biofuels Increases Greenhouse Gases through Emissions from Land-use Change." *Science* 319: 1238–40. doi: 10.1126/science.1151861.

Segev, Sigal, Maria Elena Villar, and Rosanna M. Fiske. 2012. "Understanding Opinion Leadership and Motivations to Blog: Implications for Public Relations." *Public Relations Journal* 6, no. 5: 1–31.

Selfa, Theresa, Laszloa Kulcsar, Carmen Bain, Richard Goe, and Gerad Middendorf. 2010. "Biofuels Bonanza? Exploring Community Perceptions of the Promises and Perils of Biofuels Production." *Biomass and Bioenergy* 35, no. 4: 1379–89. doi: 10.1016/j.biombioe.2010.09.008.

Sengers, Frans, Rob P. J. M. Raven, and Alex van Venrooij 2010. "From Riches to Rags: Biofuels, Media Discourses, and Resistance to Sustainable Energy Technologies." *Energy Policy* 38: 5013–27. doi: 10.1016/j.enpol.2010.04.030.

Shapouri, Hosein, James A. Duffield, and Michael S. Graboski. 1995. "Estimating the Net Energy Balance of Corn." Agricultural Economic Report No. 721. Economic Research Service, USDA, Washington, DC. https://www.ers.usda.gov/publications /pub-details/?pubid=40674.

Sharman, Amelia, and John Holmes. 2010. "Evidence-based Policy or Policy-based Evidence Gathering? Biofuels, the EU, and the 10% Target." *Environmental Policy & Governance* 20, no. 5: 309–21. doi: 10.1002/eet.543.

Sheehan, John, A. Aden, C. Riley, K. Paustian, K. Killian, J. Brenner, D. Lightle, R. Nelson, M. Walsh, and J. Cushman. 2002. *Is Ethanol from Corn Stover Sustainable? Adventures in Cyber-Farming: A Life Cycle Assessment of the Production of Ethanol from Corn Stover for Use in a Flexible Fuel Vehicle.* Golden, CO: United States Department of Energy - National Renewable Energy Laboratory. 330 pp.

Shepherd, J. Marshall. 2017. "The Hurricane Forecast 'Cone of Uncertainty' May Not Mean What You Think." *Forbes* April 17, 2017. https://www.forbes.com/sites/ marshallshepherd/2017/04/17/the-hurricane-forecast-cone-of-uncertainty-may-not -mean-what-you-think/?sh=449bac9060a0.

Shepherd, J. Marshall. 2011. "Carbon, Climate change, and Controversy." *Animal Frontiers* 1, no. 1: 5–13. doi: 10.2527/af.2011-0001.

Shrader-Frechette, Kristin. S. 2002. *Environmental Justice: Creating Equality, Reclaiming Democracy.* Environmental Ethics and Science Policy Series. New York: Oxford University Press.

Sikkema, Richard, Martin Junginger, Wilfried Pichler, Sandra Hayes, and André P. C. Faaij. 2010. "The International Logistics of Wood Pellets for Heating and Power Production in Europe: Costs, Energy-Input and Greenhouse Gas Balances of Pellet

Consumption in Italy, Sweden, and The Netherlands." *Biofuels, Bioproducts, Biorefineries* 4: 132–53. doi: 10.1002/bbb.208.

Sims, Ralph E. H. 2003. "Bioenergy to Mitigate for Climate Change and Meet the Needs of Society, the Economy and the Environment." *Mitigation and Adaptation Strategies for Global Change* 8: 349–70. doi: 10.1023/B:MITI.0000005614.51405 .ce.

Smith, Jay M. 2011. *Monsters of the Gévaudan: The Making of a Beast.* Cambridge: Harvard University Press.

Smith, Neil. 1984. *Uneven Development: Nature, Capital and the Production of Space.* Oxford: Basil Blackwell.

Smith, Neil. 1992. "Geography, Difference, and the Politics of Scale." In *Postmodernism and the Social Sciences*, edited by Joe Doherty, Elspeth Graham, and Mo Malek, 57–79. Houndmills, UK: Macmillan. doi: 10.1007/978-1-349-22183-7_4.

Smith, Jessica M., and Abraham S. D. Tidwell. 2016. "The Everyday Lives of Energy Transitions: Contested Sociotechnical Imaginaries in the American West." *Social Studies of Science* 46, no. 3: 327–50. doi: 10.1177/0306312716644534.

Smith, Slim and Andrew Hazzard. 2015. "Columbus KiOR Facility Sold to Georgia Company." *Columbus Dispatch*, October 5, 2015. http://www.cdispatch.com/news /article.asp?aid=45216.

Söderberg, Charlotta, and Katarina Eckerberg. 2013. "Rising Policy Conflicts in Europe over Bioenergy and Forestry." *Forest Policy and Economics* 33: 112–19. doi: 10.1016/j.forpol.2012.09.015.

Sorensen, Bent. 1991. "A History of Renewable Energy Technology." *Energy Policy* 19, no. 1: 8–12. doi: 10.1016/0301-4215(91)90072-V.

Sovacool, Benjamin K. 2014. "What Are We Doing Here? Analysing 15 Years of Energy Scholarship and Proposing a Social Science Research Agenda." *Energy Research and Social Science* 1: 1–29. doi: 10.1016/j.erss.2014.02.003.

Sovacool, Benjamin K., and Michael H. Dworkin. 2015. "Energy Justice: Conceptual Insights and Practical Applications." *Applied Energy* 142: 435–44. doi: 10.1016/j. apenergy.2015.01.002

Sovacool, Benjamin K., Raphael J. Heffron, Darren McCauley, and Andreas Goldthau. 2016. "Energy Decisions Reframed as Justice and Ethical Concerns." *Nature Energy* 1: 16024. doi: 10.1038/nenergy.2016.24.

Sovacool, Benjamin K., Roman V. Sidortsov, and Benjamin R. Jones. 2014. *Energy Security, Equality and Justice.* London: Routledge.

Spartz, James T., Mark Rickenbach, and Bret R. Shaw. 2015. "Public Perceptions of Bioenergy and Land Use Change: Comparing Narrative Frames of Agriculture and Forestry." *Biomass and Bioenergy* 75: 1–10. doi: 10.1016/j.biombioe.2015.01.026.

Spivak, Gayatri Chakravorty. 1993. "The Politics of Translation." In *Outside in the Teaching Machine*, edited by Gayatri Chakravorty Spivak, 179–200. London and New York: Routledge. doi: 10.4324/9780203440872.

Stack, Carol. 1996. *Call to Home: African Americans Reclaim the Rural South.* New York: Basic.

Stanturf, John A., Cees van Oosten, Daniel A. Netzer, Mark D. Coleman, and C. Jeffrey Portwood. 2001. "Ecology and Silviculture of Poplar Plantations." In *Poplar Culture in North America*, edited by Donald I. Dickmann, Judson G. Isebrands, James E. Eckenwalder, and James Richardson, 153–206. Ottawa, ON: NRC Research Press.

Statkraft. 2015. "RWE Scraps Plan to Sell Georgia Biomass." Statkraft, January 7, 2015. http://statkraftbiomass.blogspot.com/2015/01/rwe-scraps-plan-to-sell-georgia-biomass.html.

Steiger, Brad. 2012. *The Werewolf Book: The Encyclopedia of Shape-Shifting Beings.* Canton, MI: Visible Ink Press.

Stine, Jeffrey K., and Joel A. Tarr. 1998. "At the Intersection of Histories: Technology and the Environment." *Technology and Culture* 39: 601–40.

Storrow, Benjamin. 2020. "Methane Leaks Erase Some of the Climate Benefits of Natural Gas." *Scientific American* E&E News, May 5, 2020. https://www.scientificamerican.com/article/methane-leaks-erase-some-of-the-climate-benefits-of-natural-gas/.

Strauss, Claudia. 1992. "What Makes Tony Run? Schemas as Motives Reconsidered." In *Human Motives and Cultural Models*, edited by Roy D'Andrade and Claudia Strauss, 197–224. Cambridge: Cambridge University Press.

Strauss, Claudia. 2005. "Analyzing Discourse for Cultural Complexity." In *Finding Culture in Talk: A Collection of Methods*, edited by Naomi Quinn, 203–42. New York: Palgrave MacMillan.

Strauss, Claudia. 2006. "The Imaginary." *Anthropological Theory* 6: 322–44. doi: 10.1177/1463499606066891.

Strauss, Claudia. 2012. *Making Sense of Public Opinion: American Discourses about Immigration and Social Programs.* Cambridge, UK: Cambridge University Press.

Strauss, Claudia and Naomi Quinn. 1997. *A Cognitive Theory of Cultural Meaning.* Cambridge: Cambridge University Press.

Strauss, Sarah, Stephanie Rupp, and Thomas Love, editors. 2013. *Cultures of Energy: Power, Practices, Technologies.* New York: Taylor & Francis.

Susaeta, Andres, Janaki Alavalapati, Pankaj Lal, Jagannadha. R. Matta, and Evan Mercer. 2010. "Assessing Public Preferences for Forest Biomass Based Energy in the Southern United States." *Environmental Management* 45: 697–710. doi: 10.1007/s00267-010-9445-y.

Swaan, J. and S. Melin. 2008. "Wood Pellet Exports: History, Opportunities, and Challenges." PowerPoint presentation for SmallWood 2008 Conference. May 13, 2008, Madison, WI. Madison, WI: Forest Products Society. http://www.forestprod.org/smallwood08swaan.pdf.

Sweet, William V., Robert E. Kopp, Christopher P. Weaver, Jayantha Obeysekera, Radley M. Horton, E. Robert Thieler, and Chris Zervas. 2017. "Global and Regional Sea Level Rise Scenarios for the United States. NOAA [National Oceanic and Atmospheric Administration] Technical Report NOS CO-OPS 083. Silver Springs, MD: NOAA.

Sweet, Willam, Joseph Park, John Marra, Chris Zervas, and Stephen Gill. 2014. "Sea Level Rise and Nuisance Flood Frequency Changes around the United States." NOAA [National Oceanic and Atmospheric Administration] Technical Report NOS CO-OPS 073. Silver Springs, MD: NOAA.

Swyngedouw, Erik. 1997. "Neither Global nor Local: "Glocalization" and the Politics of Scale." In *Spaces of Globalization: Reasserting the Power of the Local*, edited by Kevin R. Cox, 137–66. New York: Guilford Press.

Szeman, Imre and Dominic Boyer, eds. 2017. *Energy Humanities*. Baltimore, MD: John Hopkins University Press.

Tammisola, Jussi. 2010. "Towards Much More Efficient Fuel Crops: Can Sugarcane Pave the Way?" *GM Crops* 1, no. 4: 181–98. doi: 10.4161/gmcr.1.4.13173.

Taylor, Charles. 2004. *Modern Social Imaginaries*. Durham, NC: Duke University Press.

Taylor, Dorceta. 2014. *Toxic Communities: Environmental Racism, Industrial Pollution, and Residential Mobility*. New York: New York University Press.

Tidwell, Abraham S. D., and Jessica M. Smith. 2015. "Morals, Materials, and Technoscience: The Energy Security Imaginary in the United States." *Science, Technology, and Human Values* 40, no. 5: 687–711. doi: 10.1177/0162243915577632.

Tozer, Laura, and Nicole Klenk, 2018. "Discourses of Carbon Neutrality and Imaginaries of Urban Futures." *Energy Research and Social Science* 35: 174–81. doi: 10.1016/j.erss.2017.10.017.

Trotter, Joe William, Jr., 1991. *The Great Migration in Historical Perspective: New Dimensions of Race, Class, and Gender*. Bloomington, IN: Indiana University Press.

Trutnevyte, Evelina. 2014. "The Allure of Energy Visions: Are Some Visions Better Than Others?" *Energy Strategy Reviews* 2: 211–19. doi: 10.1016/j.esr.2013.10.001.

Tsing, Anna Lowenhaupt. 2005. *Friction: An Ethnography of Global Connection*. Princeton, NJ: Princeton University Press.

Turner, James M. 2015. "Following the Pb: An Envirotechnical Approach to Lead-Acid Batteries in the United States." *Environmental History* 20, no. 1: 29–56. doi: 10.1093/envhis/emu130.

Upham, Paul, Julia Tomei, and Leonie Dendler. 2011. "Governance and Legitimacy aspects of the UK Biofuel Carbon and Sustainability Reporting System." *Energy Policy* 39, no. 5: 2669–78. doi: 10.1016/j.enpol.2011.02.036.

USDA Forest Service. 2016. "The Southern Forest Products Industry." https://usfs .maps.arcgis.com/apps/MapSeries/index.html?appid=7f8429df087e4c86951a7e6 9d93207a7.

US DOE. 2006. "Breaking the Biological Barriers to Cellulosic Ethanol: A Joint Research Agenda." Report from the December 2005 Workshop, DOE/SC-0095. U.S. Department of Energy Office of Science. www.genomicscience.energy.gov /biofuels/.

US DOE. 2011. *U.S. Billion-Ton Update: Biomass Supply for a Bioenergy and Bioproducts Industry*. Richard D. Perlack and Bryce J. Stokes (leads). ORNL/ TM-2011/224. Oak Ridge, TN: Oak Ridge National Laboratory. https://www .energy.gov/eere/bioenergy/downloads/us-billion-ton-update-biomass-supply-bio-energy-and-bioproducts-industry.

US DOE. 2016. *2016 Billion-Ton Report: Advancing Domestic Resources for a Thriving Bioeconomy, Volume 1: Economic Availability of Feedstocks*. Matthew H. Langholtz, Bryce J. Stokes, and Laurence M. Eaton (leads), ORNL/TM-2016/160.

Oak Ridge, TN: Oak Ridge National Laboratory. 448p. https://doi.org.10.2172 /1271651. http://energy.gov/eere/bioenergy/2016-billion-ton-report.

van De Graaf, Thijs, and Benjamin K. Sovacool. 2020. *Global Energy Politics.* Cambridge, UK: Polity Press.

van der Horst, Dan, and Saskia Vermeylen. 2011. "Spatial Scale and Social Impacts of Biofuel Production." *Biomass and Bioenergy* 35, no. 6: 2435–43. doi: 10.1016/j. biombioe.2010.11.029.

van Vuuren, Detlef P., Kaj-Ivar van der Wijst, Stijn Marsman, Maarten van den Berg, Andries F. Hof, and Chris D. Jones. 2020. "The Costs of Achieving Climate Targets and the Sources of Uncertainty." *Nature Climate Change* 10: 329–34. doi: 10.1038/s41558-020-0732-1.

Vedwan Neeraj, and Robert E. Rhoades. 2001. "Climate Change in the Western Himalayas of India: A Study of Local Perception and Response." *Climate Research* 198: 100–17. doi: 10.3354/CR019109.

Vel, Jacqueline. 2014. "Trading in Discursive Commodities: Biofuel Brokers' Roles in Perpetuating the Jatropha Hype in Indonesia." *Sustainability* 6: 2802–21. doi: 10.3390/su6052802.

Venuti, Lawrence, ed. *The Translation Reader*, 3rd Edition. 2000. London: Routledge.

Vitosh, M. L., R. E. Lucas, and G. H. Silva. 1997. "Long-term Effects of Fertilizer and Manure on Corn Yield, Soil Carbon, and Other Soil Chemical Properties in Michigan." In *Soil Organic Matter in Temperate Agroecosystems,* edited by Eldor A. Paul, Keith Paustian, E. T. Elliot, and C. V. Cole, 129–39. Boca Raton, FL: CRC Press.

Vogel, Kenneth P. 1996. "Energy Production from Forages (or American Agriculture – Back to the Future)." *Journal of Soil and Water Conservation* 51: 137–39.

Vogel, Kenneth P. 2004. "Switchgrass." In *Warm-season (C4) Grasses*, edited by Moser, Lowell E., Lynn Sollenberger, and Byron Burson, 561–88. ASA-CSSA-SSSA Monograph. Madison, WI.

Wagner, James A. 2008. "Buying Oil from People Who Hate Us." Letter to the Editor, *Chicago Daily Herald*, 1 August 2008. http://prev.dailyherald.com/story/ ?id=224758.

Walker, Gordon, Neil Simcock, and Rosie Day. 2016. "Necessary Energy Uses and a Minimum Standard of Living in the United Kingdom: Energy Justice or Escalating Expectations?" *Energy Research & Social Science* 18: 129–38. doi: 10.1016/j. erss.2016.02.007.

Walker, Thomas, Peter Cardellichio, Andrea Colnes, John Gunn, Brian Kittler, Bob Perschel, Christopher Recchia, and David Saah. 2010. *Biomass Sustainability and Carbon Policy Study*. Brunswick, ME: Manomet Center for Conservation Sciences.

Wang, Michael, May Wu, and Hong Huo. 2007. "Life-cycle Energy and Greenhouse Gas Emission Impacts of Different Corn Ethanol Plant Types." *Environmental Research Letters* 2: doi: 10.1088/1748-9326/2/2/024001.

Watts, Jonathon. 2021. "Climatologist Michael E Mann: 'Good People Fall Victim to Doomism. I Do Too Sometimes'." *The Guardian* February 27, 2021. https://www .theguardian.com/environment/2021/feb/27/climatologist-michael-e-mann-doom-ism-climate-crisis-interview.

Watts, Laura. 2018. *Energy at the End of the World: An Orkney Islands Saga.* Cambridge, MA: The MIT Press.

Wear, David N. 2002. "Land Use." In *Southern Forest Resource Assessment*, edited by David N. Wear and John G. Greis, 153–173. General Technical Report SRS-53. Asheville, NC: U.S. Department of Agriculture Forest Service Southern Research Station.

Wear, David, Robert Abt, Janaki Alavalapati, Greg Comatas, Mike Countess, and Will McDow. 2010. "The South's Outlook for Sustainable Forest Bioenergy and Biofuels Production." *Bioenergy* 1–20. http://www.treesearch.fs.fed.us/pubs /36461.

Wear, David R., and John G. Greis. 2002. "Southern Forest Resource Assessment." Gen. Tech. Rep. SRS-53. Asheville, NC: USDA Forest Service, Southern Research Station.

Wear David N., Jeffrey Prestemon, Robert Huggett, and Douglas Carter. 2013. "Markets." In *The Southern Forest Futures Project: Technical Report*, edited by David N. Wear and John G. Greis, pp 183–212. General Technical Report SRS-178. Asheville, NC: U.S. Department of Agriculture Forest Service, Southern Research Station.

Wesselink, Anna, Karen S.Buchanan, Yola Georgiadou, and Esther Turnhout. 2013. "Technical Knowledge, Discursive Spaces and Politics at the Science–Policy Interface." *Environmental Science & Policy* 30: 1–9. doi: 10.1016/j. envsci.2012.12.008.

White House. 2021. "Executive Order on Tackling the Climate Crisis at Home and Abroad." https://www.whitehouse.gov/briefing-room/presidential-actions/2021/01 /27/executive-order-on-tackling-the-climate-crisis-at-home-and-abroad/.

Whitington, Jerome. 2019. *Anthropogenic Rivers: The Production of Uncertainty in Lao Hydropower.* Ithaca, NY and London: Cornell University Press.

Whyte, Kyle Powys. 2011. "The Recognition Dimensions of Environmental Justice in Indian Country." *Environmental Justice* 4 (4): 199–205. doi: 10.1089/ env.2011.0036.

Wicker, Gerald. 2002. "Motivation for Private Forest Landowners." In *Southern Forest Resource Assessment,* edited by David N. Wear and John G. Greis, 225–37. General Technical Report SRS-53. Asheville, NC: U.S. Department of Agriculture Forest Service, Southern Research Station.

Wilhelm, Wallace, W., J. M. F. Johnson, Jerry L. Hatfield, W. B. Voorhees, and D. R. Linden. 2004. "Crop and Soil Productivity Response to Corn Residue Removal: A Literature Review." *Agronomy Journal* 96: 1–17. doi: 10.2134/ agronj2004.0001.

Wilkinson, Mike, and Mark Tepfer. 2009. "Fitness and Beyond: Preparing for the Arrival of GM crops with Ecologically Important Novel Characters." *Environmental Biosafety Resources* 8: 1–14. doi: 10.1051/ebr/2009003.

Witter, Rebecca, Kimberly R. Marion Suiseeya, Rebecca L. Gruby, Sarah Hitchner, Edward M. Maclin, Maggie Bourque, and J. Peter Brosius. 2015. "Moments of Influence in Global Environmental Governance." *Environmental Politics* 24, no. 6: 894–912. doi: 10.1080/09644016.2015.1060036.

Wolsink, Maarten. 2007. "Wind Power Implementation: The Nature of Public Attitudes: Equity and Fairness Instead of 'Backyard Motives'." *Renewable and Sustainable Energy Reviews* 11, no. 6: 1188–207. doi: 10.1016/j.rser.2005.10.005.

Wood, Spencer D. and Jess Gilbert. 2000. "Returning African American Farmers to the Land: Recent Trends and a Policy Rationale." *The Review of Black Political Economy* Spring 2000: 43–64. doi: 10.1007/BF02717262.

Woodwork Institute. 2004. "Green Comparison: SFI and FSC." *Archetype* Fall/ Winter 2004: 10–11.

Wright, Wynne, and Taylor Reid. 2011. "Green Dreams or Pipe Dreams?: Media Framing of the U.S. Biofuels Movement." *Biomass and Bioenergy* 35: 1390–99. doi: 10.1016/j.biombioe.2010.07.020.

Yaman, Serdar. 2004. "Pyrolysis of Biomass to Produce Fuels and Chemical Feedstocks." *Energy Conversion and Management* 45: 651–67. doi: 10.1016/ S0196-8904(03)00177-8.

Yan, Yuzhen, Michael L. Bender, Edward J. Brook, Heather M. Clifford, Preston C. Kemeny, Andrei V. Kurbatov, Sean Mackay, Paul A. Mayewski, Jessica Ng, Jeffrey P. Severinghaus, and John A. Higgins. 2019. "Two-Million-Year-Old Snapshots of Atmospheric Gases from Antarctic Ice." *Nature* 574: 663–66. doi: 10.1038/s41586-019-1692-3.

Zabawa, Robert. 1991. "The Black Farmer and Land in South Central Alabama: Strategies for Preserving a Scarce Resource." *Human Ecology* 19, no. 1: 61–80. doi: 10.1007/BF00888977.

Zabawa, Robert, Arthur. Siaway, and Ntam Baharanyi. 1990. "The Decline of Black Farmers and Strategies for Survival." *Southern Rural Sociology* 7: 106–21. https:// egrove.olemiss.edu/jrss/vol07/iss1/9.

Zhang, Daowei, Brett J. Butler, and Rao V. Nagubadi. 2012. "Institutional Timberland Ownership in the US South: Magnitude, Location, Dynamics, and Management." *Journal of Forestry* 110, no. 7: 355–61. doi: 10.5849/jof.12-015.

Index

Alabama, 27, 29, 33, 53, 78, 177
Al Gore, 3, 142, 149
anthropology, xii, 4, 20, 23, 25, 56, 68, 130, 193

Billion Ton Reports, 47
bioenergy conferences, 4, 27, 28, 39, 58, 65, 120, 137, 143, 144
bioenergy industry, 2, 3, 6, 15, 28, 30, 38, 40, 44, 64, 69, 75, 95, 98, 119, 122, 137, 152, 170
bioenergy with carbon capture and storage (BCCS), 11, 135
biomass, woody, 6, 8–9, 13, 16, 24, 38, 40, 43, 60, 64–66, 75–76, 84–85, 90, 92, 99, 105, 111, 119, 134–35, 138, 174, 187–88
biomass fluid catalytic cracking (BFCC), 32, 41
Black landowners: Black forest management, 4, 15, 26, 152–54, 173–74; Black landowners, 4, 14–15, 26, 151–53, 170–74, 179; Black-owned land, 4, 14, 160, 173

carbon, 11–13, 17, 37, 42, 50, 52, 55, 70–71, 115, 119–20, 122–25, 133, 135, 137, 139, 143–47, 183, 189, 191; carbon credits, also

carbon markets, 146; carbon life cycle analysis, 37, 42, 145; carbon neutrality, 11, 13, 17, 37, 39, 70, 119, 137, 145, 147; carbon sequestration, 42, 55, 122, 146
carbon capture and storage (CCS), 11. *See also* bioenergy with carbon capture and storage (BCCS)
carbon credits, also carbon markets, 146
carbon life cycle analysis, 37, 42, 145
carbon neutrality, 11, 13, 17, 37, 39, 70, 119, 137, 145, 147
carbon sequestration, 42, 55, 122, 146
cellulosic biocrude, 4, 6, 29, 30, 32, 40–41, 47, 49, 54, 93
cellulosic ethanol, 1, 2, 4, 6, 12, 24, 29–30, 40–41, 47, 62, 93–95, 98, 102, 115
certification: certification, 48, 89, 90, 120, 138–40; FSC, 48, 138; Sustainable Forestry Initiative (SFI), 48, 90, 138–39; Tree Farm, 48
civil rights, xi, 160–64, 185
Civil War, xi, 7, 158
climate change: adaptation, 11, 35, 127, 134; anthropogenic climate change, 13, 121, 123, 125, 127–28, 133, 135, 193; climate fluctuations or natural variability, 11, 35, 44, 124–25, 133;

231

rural development, 3, 6, 7, 15–17, 29,
37, 39, 43–45, 47–48, 52–53, 55,
59, 61, 68, 76, 92–93, 98, 100–112,
120–22, 136, 148, 152, 187–88, 191

scale, xii, 2, 5, 7, 9, 11–13, 16, 18–20,
23, 25–26, 28–32, 38–43, 47, 52–53,
57–58, 73, 75–76, 79, 81–82, 85, 95,
98, 116, 122, 128, 135, 143, 147–48,
151, 154, 157, 169, 181, 184, 187,
189, 193; multi-scalar, 18, 23, 29,
34, 51, 96, 111, 183, 190. See also
politics of scale
science and technology studies (STS),
49, 121
segregation, 7, 160–61, 163, 170, 172
sharecropping, 7, 14, 157–58, 185n2
silviculture, 1, 75, 96, 114, 116, 189
siting of bioenergy facilities, 15, 17,
25, 29–30, 50, 54, 70, 76–77, 99,
104–05, 110, 148, 152, 174–76, 178,
180, 184
slavery, 7, 9, 14, 78, 151, 157–59, 161,
166, 170–71; enslaved persons, 151,
156–59
social acceptability of bioenergy, 24,
25, 35n7, 44, 48, 50, 181; protests or
opposition, 16, 17, 44, 52–53, 70, 76,
85, 94, 104–05, 111, 146–47, 176,
181–82
sociotechnical imaginaries, 16, 24, 45,
49, 50, 51, 53, 56–57, 110, 134, 147
Soperton, Georgia, 1, 2, 10, 19, 24, 28–
31, 62, 77, 79, 82, 88, 93, 95–101,
105–6, 108–11, 167, 181, 192
subsidies, 6, 17, 19–20, 23, 43, 46, 61,
77, 89, 94–95, 113–14, 119, 166, 191

sustainability, 3, 6, 7, 12–13, 34, 39, 42,
44, 48, 55, 59, 65–66, 77, 85, 89–90,
120–22, 138–40, 145, 147, 169, 191,
194

taxpayers, 2, 8, 30, 61–62, 68, 96, 103,
105, 108
technological innovation, 32, 37, 50, 52,
55, 61, 95, 191
trade-offs *vs.* win-win, 42, 61, 100, 169
translation, 3, 5, 16, 20–23, 38, 52, 69,
77, 111, 122, 125, 148, 154, 163,
187–89, 192
transportation, xi, 6, 9, 11, 30, 42, 46–
47, 71n1, 92, 101, 122, 148, 192
turpentine, 1, 73, 75, 80, 96, 108, 152,
158–60, 167

UNCED (United Nations Conference on
Environment and Development), also
called Earth Summit or Rio Summit,
128
UNFCCC (United Nations Framework
Convention on Climate Change),
128–29, 192
United Kingdom (U.K.), 12
United Nations, 126, 128

valuation, of forest and land, 9, 14, 77–
84, 113, 115, 152–53, 171–74, 193

waste wood and forest residues, 2, 10,
13, 20, 40–41, 44, 87–89, 98–100,
103, 140, 170
welfare, 166–67
White privilege, 34, 100, 153, 161,
163–64

About the Authors

Sarah Hitchner is an assistant research scientist and adjunct professor in the Department of Anthropology at the University of Georgia. She is a cultural anthropologist who has published widely on African American landowners, heirs' property, cultural landscapes, Malaysian and Indonesian studies, the southeastern United States, ecotourism, conservation, and bioenergy. Her dissertation research focused on participatory mapping of the highly anthropogenic cultural landscape of the Kelabit Highlands in Sarawak, Malaysia. She is co-director of the UGA Bali & Beyond study abroad program in Bali, Indonesia, where she teaches Ethnographic Writing. She is coeditor (with Fausto Sarmiento) of *Indigeneity and the Sacred: Indigenous Revival and the Conservation of Sacred Natural Sites in the Americas* (Berghahn Books, 2017) and coeditor (with J. Peter Brosius) of the forthcoming *Handbook on Conservation and Social Science* (Edward Elgar Publishing, 2023).

John Schelhas is a research forester with the Southern Research Station of the USDA Forest Service in Athens, Georgia. His research centers on relationships between people and forests, His primary research focus is private forest landowners and rural communities, addressing topics such as land use decision-making, cultural models and imaginaries of forests, forest-based rural development, race and ethnicity, relationships between protected areas and their neighbors, invasive plants, bioenergy, and climate change. He has conducted research in Latin America and the U.S. South. He holds a PhD in Renewable Natural Resources with a Minor in Anthropology from the University of Arizona. He is the coauthor (with Max J. Pfeffer) of *Saving Forests, Protecting People? Environmental Conservation in Central America* (AltaMira Press, 2009) and coeditor (with Louise E. Buck, Charles Geisler, and Eva Wollenberg) of *Biological Diversity: Balancing Interests through*

Adaptive Collaborative Management (CRC Press, 2001) coeditor (with Russell Greenberg) of *Forest Patches in Tropical Landscapes* (Island Press, 1996).

J. Peter Brosius is a distinguished research professor of anthropology at the University of Georgia and founding director of UGA's Center for Integrative Conservation Research. He received his M.A. from the University of Hawai'i and his PhD from the University of Michigan. He is widely recognized for his work with Penan hunter-gatherers in Sarawak, Malaysia, and for his contributions to the development of political ecology. Throughout his career, he has been engaged with issues of environmental degradation, indigenous rights and conservation. His research in Malaysia focused on the international rainforest campaign centered on Penan, and he continues to be involved in court testimony in support of Penan land claims against oil palm companies. He also played a formative role in the development of Collaborative Event Ethnographies of major conservation meetings. In addition to his research on bioenergy in the American South, he has ongoing research on the Tolak Reklamasi movement in Bali, Indonesia, and the political ecology of real estate in occupied surfscapes in the Global South. Brosius is author of the book *After Duwagan: Deforestation, Succession and Adaptation in Upland Luzon, Philippines* (Center for South and Southeast Asian Studies, University of Michigan, 1990), coeditor (with Anna Tsing and Charles Zerner) of *Communities and Conservation: Histories and Politics of Community-based Natural Resource Management* (Altamira Press, 2005), and coeditor (with Sarah Hitchner) of the forthcoming *Handbook on Conservation and Social Science* (Edward Elgar Publishing, 2023).

www.ingramcontent.com/pod-product-compliance
Lightning Source LLC
Chambersburg PA
CBHW050639280326

41932CB00015B/2707